The Immaterial Society

THE IMMATERIAL SOCIETY

Design, Culture, and Technology in the Postmodern World

Marco Diani
Editor

Northwestern University
Centre Nationale de la Recherche Scientifique

PRENTICE HALL
Englewood Cliffs, New Jersey 07632

Library of Congress Cataloging-in-Publication Data

The Immaterial Society : design, culture, and technology in the postmodern world / Marco Diani, editor.
p. cm.
Includes bibliographical references and index.
ISBN 0-13-203027-6
1. Postmodernism. 2. Technology. 3. Civilization, Modern—1950–
I. Diani, Marco.
BH301.P69I46 1992
306.4′6—dc20 91-32800
CIP

Acquisitions editor: Nancy Roberts
Production editor: Merrill Peterson
Copy editor: Rene Lynch
Cover designer: Rich Dombrowski
Indexer: Susan Coerr
Prepress buyer: Kelly Behr
Manufacturing buyer: Mary Ann Gloriande
Editorial assistant: Pat Naturale

Printed in the United States of America

10 9 8 7 6 5 4 3 2 1

ISBN 0-13-203027-6

Prentice-Hall International (UK) Limited, *London*
Prentice-Hall of Australia Pty. Limited, *Sydney*
Prentice-Hall Canada Inc., *Toronto*
Prentice-Hall Hispanoamericana, S.A., *Mexico*
Prentice-Hall of India Private Limited, *New Delhi*
Prentice-Hall of Japan, Inc., *Tokyo*
Simon & Schuster Asia Pte. Ltd., *Singapore*
Editora Prentice-Hall do Brasil, Ltda., *Rio de Janeiro*

Contents

Preface

After the "great debates" on the nature and the consequences of technology, the role of design appears to be destined to dramatically increase in the next few years, expanding its importance and thus requiring some substantial changes also in the nature of design.

In *The Immaterial Society*, I have gathered together for the first time a body of new writing in the emerging field of "design studies." The contributors to the book argue in different ways and from different intellectual standpoints for a rethinking of design in light of its cultural significance and its powerful position in today's society. The essays assembled in the following chapters address such topics as the relations among design, technology, and social organization; the "intellectual historiography" or "genealogy" of the concepts of postmodern and immaterial; the applications of rhetoric, semiotics, and phenomenology in interpreting and analyzing design. Drawing on postmodernist theory as well as other critical strategies, the writers also show how design has become a complex and multidisciplinary activity, providing startling and provocative ideas on the future of design, insights from architecture, philosophy, sociology, art criticism, literary theory, and cultural history by some of the key thinkers of our time.

One of the main characteristics of the dominant structural form of work in post-industrial societies is the pervasiveness and the adaptability of new technologies. Fast technological change from human or machine-mediated work to computer-mediated work leads to changes in how individuals and groups adapt to their particular work environment. Concomitant with these technological shifts are societal and cultural transformations—from industrial to post-industrial societies—reflecting changes from a society based on manufacturing and producing material goods to a society based on services, that is, nonmaterial goods.

Under these new conditions, *design* becomes a more complex and multidisciplinary operation than in the recent past: What is, in fact, to be designed? Chairs, screens, or keyboards, as "traditional" designers continue to do? Spaces and buildings, as architects have been doing increasingly in the last ten years, proclaiming, through the "postmodernist" vulgate, their farewell to the remnants of our industrial societies? Who is going to decide on the content, mode of access, and operation of programs and procedures which are the basis for so many complex organizations? What kind of competences are going to be required to design new working spaces, process, and products which are eminently *cognitive* and *immaterial*?

These questions call for a "new design" for the Immaterial Society. Resulting from interdisciplinary research, with contributions stemming from many different disciplines, all these essays, by some of the key thinkers of our time, define the "architecture of complexity" which is the most important and vulnerable element of a society dominated by computer-mediated work.

The introductory text places these essays, all of which appeared in a special issue of the journal *Design Issues*, in context, and the closing bibliographical essay provides a significant account of the connections between design, culture, and society.

Interdisciplinary research is an exercise more widely invoked than practiced and, for many reasons, understandably so. It is never easy, in fact, to abandon well-established and convenient intellectual settings and boundaries or to face questions that arise from both theory and practice, coming from elsewhere in time, space, and culture.

Interdisciplinarity upsets established routines and blurs accustomed visions of self and of other: Newcomers and outsiders—almost synonymous in this case—subvert the "order of things." One essential condition for interdisciplinarity is that, in an early stage of the process, the "laws of hospitality" work extremely well. I am pleased to have been the witness and to certify that such was the case for the preparation of this book.

The task could never have been completed without the open-minded hospitality, the patience, and the intellectual support of those who worked with the "guest": All members of the editorial board of *Design Issues* have been supportive, even when the project seemed to explore areas and disciplines at first sight far remote from their own interests. In other cases such exploration would have led, almost inevitably, to disagreement and friction; in ours it deepened intellectual debates and cemented friendships. *Primus inter pares*, Martin Hurtig has been the "skeptical stimulator" who kept the project going, and going somewhere, while the School of Art and Design has been very supportive, materially and intellectually. Martha D. Pollak, Department of History of Art and Architecture, the University of Illinois at Chicago, provided very valuable intellectual criticism and guidance.

Also, I would have been constantly lost in technical and editorial intricacies without the help and the patience of Faith Van Alten and Bonnie Osborne, whose courtesy and competence never failed, in spite of the difficulties of dealing with a peripatetic, or rather, nomadic, guest editor. In a way, this position and its condition suited this interdisciplinary project. I also would like to thank my editor at Prentice Hall, Nancy Roberts, and Merrill Peterson, my production editor.

Last, but not least, the support that stemmed from various institutions for this and related projects must be acknowledged:

- The Italian Cultural Institute of Chicago and its director, Professor Silvio Marchetti, readily provided support, both material and intellectual, in particular for the translation of some essays.
- The Center for Interdisciplinary Research in the Arts (CIRA) at Northwestern University and, in particular, its director Susan Lee not only provided me with a research fellowship but helped in many other ways. For instance, some of the questions addressed in these pages were also the object of a conference on "Design, Technology, and the Future of Post-Industrial Societies," sponsored by the CIRA and co-organized with Victor Margolin and Larry Salomon. In that connection, Angela Murphy provided very valuable help.
- Year after year, the *Centre National de la Recherche Scientifique* (CNRS) in Paris has provided what is probably the most invaluable resource of all: the freedom to undertake and pursue projects during a long period. This characteristic makes it an almost unique institution in the world of knowledge and of real interdisciplinary research, and I am pleased once again to acknowledge my gratitude for being part of it.

About the Authors

THE EDITOR

Marco Diani, sociologist, is *Chargé de recherche* at the prestigious *Centre Nationale de la Recherche Scientifique* (CNRS) in Paris, where he works with Alain Touraine. He was also, from 1985 to 1990, Visiting Professor of Sociology and French and Italian Studies at Northwestern University. Author or co-author of more than 80 publications, he has published extensively in the area of new technologies and organizational change, sociology of work, and contemporary social theory. His books include *Restructuring Architectural Theory* (1989) and *L'intelligenza dell'automazione* (1991).

CONTRIBUTORS

Maurizio Ferraris is Professor of Esthetics at the University of Trieste and editor of the journal *Aut-Aut*. He is the author of several books on critical theory, deconstruction, and literature, such as: *Storia dell'ermeneutica* (1989).

Fred Forest, one of the pioneers of video art, is Professor at the *Ecole Nationale d'Art de Cergy* and adjunct professor at La Sorbonne, University of Paris. Among his publications are *Art Sociologique/Vidéo*, and numerous essays in journals of art, theory, and sociology. He has lectured widely and participated in the *Biennale* in Vienna and Sao Paulo.

Omar C. Iadin is the Director of the *Centre de Recherches sur les Innovations* in Paris. He is the author of *Déplacements quotidiens de main d'oeuvre et conflits dans la région milanaise* (1978) and of numerous publications about the social impact of innovation and has lectured widely on the subject throughout the world.

Jeff Kintzelé is a sociologist at the *Centre d'Etude de la Société Luxembourgeoise* and was, from 1985 to 1989, a researcher at the European University Institute in Florence. He is author of a sociological analysis of the flirt, *La drague* (1989) and is currently working on a book about the sociology of money.

Michel Maffesoli is Professor of Sociology at La Sorbonne, University of Paris, and Director of the *Centre d'Etude sur l'Actuel et le Quotidien* (CEAQ). The editor of the international journal **Sociétés**, he has written several books, translated in many languages, among them *L'ombre de Dionysos: Contribution à une sociologie de l'orgie* (1985); *La connaissance ordinaire* (1985); and *Le temps des tribus* (1987).

Victor Margolin, Associate Professor of Design History at the University of Illinois at Chicago, was the co-founder and first editor of the international journal *Design Issues*. The author of several books and numerous articles on art and design history, he has recently published *Design Discourse: History Theory Criticism* (1989).

Abraham A. Moles is Professor of Sociology and Social Psychology at the *Université Louis Pasteur* and Director of the Institute of Social Psychology of Communications, both in Strasbourg. He is the author of more than 500 publications and twenty books, translated in many languages, among which are *Theory of Acts: Toward an Ecology of Action* (1977) and *Psychology of Space* (1982). His most recent book is *Les sciences de l'imprécis* (1991).

Maurizio Morgantini is an architect, planner, and industrial designer working in Milan. From 1984 to 1990, he held the chair of Industrial Design at The University of Illinois at Chicago. His work has been widely published. He researches the hyperartificial condition of mankind with reference to the new technologies and he has lectured widely on the subject throughout the world.

Tufan Orel is Research Associate and Lecturer at the *Université Technologique de Compiègne*, France, where he teaches sociology and history of science. He has published several articles in the field of methodology of social sciences and philosophy. His main research interest is the theories of human action and their application to the postindustrial society.

Victor Scardigli is *Directeur de recherche* at the *Centre National de la Recherche Scientifique* (CNRS) in Paris, where he heads a research team on everyday life, communications, and development. Among his publications are *La société digitale* (1987) and *La consommation culture du quotidien* (1984).

Daniel Simeoni is a research associate with the University of Paris-VII, Jussieu. He has published several articles on the subject of interrelations between language and images.

Herbert A. Simon, the 1978 Nobel Prize Laureate for Economics, is Richard King Mellon Professor at Carnegie-Mellon University and is one of the most important figures in the field of artificial intelligence. He has authored numerous books and publications which have been widely translated, among which are *The Sciences of the Artificial* (1968); *Models of Thought* (1969); and a recent autobiography, *Models of My Life* (1990).

Norbert Wiener (1894–1964), the discoverer of cybernetics, was Institute Professor of Mathematics at the Massachusetts Institute of Technology until his death and was the author of numerous books and publications which have been widely translated. Among his most important are *Cybernetics* (1961); *The Human Use of Human Beings* (1954); and a remarkable autobiography in two volumes.

The Immaterial Society

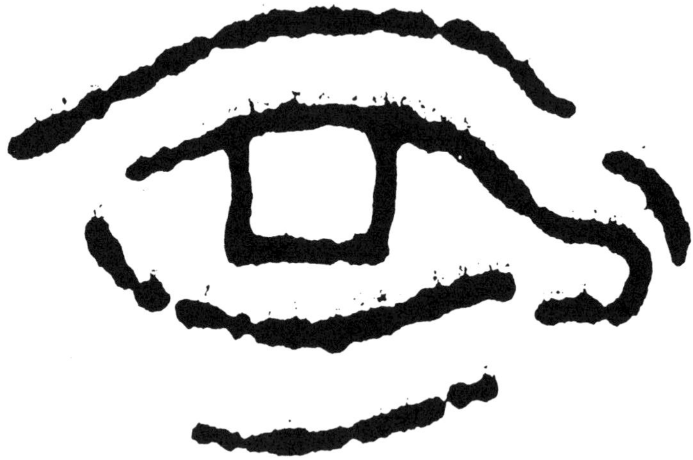

FIGURE 1. Leonardo da Vinci, *Scritti letterari di Leonardo da Vinci*, ed. Jean Paul Richter, Sampson Low, Murston, Searle, and Rivington, London, 1883 (reprinted by Phaidon Press, London, 1970).

Immateriality Takes Command[1]

Marco Diani

DESIGN, TECHNOLOGY, AND THE FUTURE OF POSTINDUSTRIAL SOCIETY[2]

If it is true that this is a watershed period for the world, born more than two centuries ago with the industrial revolution, then *industrial design* (one should never forget its origins) is an essential part of the new digital transformation. Once this is said, one still has to justify the reasons and the motivations behind the enterprise the reader holds in hand.

The ideal introduction would look like the book Schopenauer dreamed of, where one could read eight or more stories at the same time, or even the unique book written by Raymond Queneau—hundreds and hundreds of self-contained statements that can also be read in association with all the others. So it goes from the haiku to the *opera aperta* in a single text through an almost infinite number of combinations and possibilities.[3] In the face of such complex phenomena as the ones examined in the following chapters, aphorisms would be better suited to introduce the debates going on in the field of design than any schematic attempt to summarize.[4]

This multidisciplinary ensemble is to be considered as an introduction to the understanding of the new aspects of design in the transitional period from the industrial and modern age to the immaterial and postmodern age.[5] The fall of the logical principle of general coherence is one of the characteristics of our times: simultaneous presence of opposed, contradictory phenomena, exemplified in certain facts, such as the growing automation of man–machine relations of everyday life in general, and at the same time, research in design oriented to the enhancement of purposeless lyrical values, such as Mendini's "poetic reaction objects."[6]

Even if we disagree with the terminology, it is difficult to deny that, at least in our societies, we are witnessing the end of most of the traditional oppositions—for instance, material–immaterial, mind–body, terrestrial–celestial, subjectivism–individualism—that contributed so much to the framing of our mode of understanding.

By the same token, are we really in the age of the crisis of reason, of *Krisis*, the age of impossibility of general norms and methods? Or are we more sensitive and prepared to perceive and analyze reality as a multipolar, multicentric phenomenon? The investigations gathered in this book all deal with this dominant set of questions. Dominant questions that are going to dominate us? In the future? Now?

Where do we start? From that recurring intellectual problem that was defined more than twenty years ago by Herbert A. Simon: If the sciences of the artificial are opposed to the natural sciences, design is the scientific method to interpret and solve the problems related to the "architecture of complexity." Some argue or have argued (probably less today) that design should be considered more a technological or an artistic activity than a scientific one. The question seems so intricate and complex that, instead of giving just one answer (the *one best way*) we decided to also give the reader some historical foundations about what could aptly be termed the science and conscience of design.[7]

Is design a science? Can there be a science of design?[8] Many attempts have been aimed at creating a scientific environment for design or, to put it rather bluntly, to create a science of design, a design science. This task has not been easy, yet it has created noticeable concern in many design conferences and publications during the past twenty years.

The basic text that launched the discussion and the search for a science of design is undoubtedly Herbert A. Simon's *The Sciences of the Artificial*: It is indeed impressive to see how influential such a slim volume has been in the debates about design method and scientific method.[9] Professor Simon's accomplishments and publications are too numerous to be listed here,[10] and there is no need to introduce further explanations into his argument. Only one point should be stressed: If "the natural sciences are concerned with things as they are," Simon explains in a paradoxical way,[11] "design is concerned with how things ought to be." At first sight, then, there is a distinction between science and design; in fact, Simon outlines a series of characteristics that would aptly represent one of the "sciences of the artificial" design: "a body of intellectually tough, analytical, partly formalizable, partly empirical, teachable doctrine about the design process."

Through the life, actions, and writings of Norbert Wiener, we can analyze one of the most important cases of "scientific critique" of science and technology. His fame is based both on his scientific achievements and his independent and strongly polemic position on the use and total control of science and technology by political power. He can be likened to the kind of intellectual who asserts that "whether or not it draws on new research, technology is a branch of moral philosophy," described by Paul Goodman. In fact, a preoccupation of Wiener's since the early 1940s was his working toward the construction of a practical philosophy of technology for the modern age, a field nearly unworked hitherto but one that made high demands on the thinker in order to break through the resistance offered by conventional wisdom, common sense, and apologetics for vested interests that surrounded it.

Why and when has the legitimacy of science and technology been called into question by the scientists themselves? Criticism of science and technology had a relatively minor development until World War II, when faith in the identity of scientific, technological, and social progress remained largely unshaken in the Western industrialized countries. Even during and after the Great Depression of the 1930s, most people, as well as scientists and governments, regarded the enhancing value of scientific knowledge and its technological implications as self-evident. But since Hiroshima and the dominant positioning of military-science-based industry, public anxiety about the consequences of scientific discoveries has risen steadily. The nuclear arms race, the polluting and carcinogenic effects of new petrochemicals, the actual and possible use of electronic devices as instruments of social control, the prominent part played by certain science-based technologies in multiple local wars, the potentialities for genetic engineering created by the advances in molecular biology, and the implications of automation on employment and the workplace: These are only some conspicuous causes for the rising alarm about the result, the use, and the manipulation of scientific research and technology.

In the past twenty years, automation, computerization, robotics, and so forth have become passwords too widely used in societies involved in continuing and pursuing the "next industrial revolution." Notwithstanding the growing mass of literature in the field, it is very difficult therefore to find what to read to understand our present condition in a world increasingly dominated by the power of science, technology, and industry in every aspect of life. This is the technological veil that was well described by David Noble in his *America by Design*:

> [Because] as technology has increasingly placed the world at people's fingertips, these people have become less able to put their finger on precisely what technology is. A general mystification evolved just as modern technology was becoming a dominant aspect of social life. Paradoxically, perhaps, the popular perception of technology as an autonomous prime mover in history, independent of human will, became widespread at precisely the time when the social process that is modern technology in fact began to come under the conscious control of human authority in the specific form of private corporate capital.

Does anyone remember the question of the "two cultures" (the scientific and the literary) much debated all over the world during the 1950s? Judging from many of the theoretical convergences of the essays contained in this book, it seems there is a growing area somewhat similar to a no man's land in which design could become the essential, required link, or third factor, between the unilateral growth of science and technology and the human and cultural complexity of postindustrial society.[12]

A major phenomenon that we have observed in recent years has a greater importance and impact on our lives than we previously realized: the question of communication.[13] Emphasis is more and more on languages and their interdependence and relation to the complexity of social life: from deconstruction of the modern project[14] to the possibility of giving new meanings to objects as cultural and cognitive processes,[15] with computers playing a central role in this "digitalization of mankind."[16]

New horizons appear from the standpoint of "computer-mediated life" and, after the great debates on the nature and the consequences of technology, the role of design will dramatically increase in the next few years, expanding its importance and thus requiring some substantial changes in the nature of design.

Artificiality or the ecstasy of the unnatural: Hasn't the world of design always been the world of artifacts? Is it an artificial world? Not only do we intend to acknowledge the terminology used by Iadin in his contribution but also to make a plea for a reassessment of the relations between functionality and materiality in the field of design. In order to do that, we do not have to use the latest theoretical debates in postmodernist theory but rather to explore the numerous roads not taken in the development of industrial design.

Arnold Toynbee, who is certainly not to be suspected of revolutionary, subversive views, once wrote that humans transform inanimate raw materials into tools and give them a function and a style that raw materials did not have before. Functions and styles are immaterial: In material terms they are created *ex nihilo*.[17]

Why then consider immateriality as a new radical aspect of our reality? Is it because emphasis is now more and more on language and its consequences? Design and the immaterial society: The terminology is not entirely new. Research goes on in various countries. Design has always been torn between two poles of our dominant culture: technology and human scale productivity, industrial reality and social utopia. An old battle, with *romantic* opposed to *rationalist*, is playing again with postmodernist opposed to modernist. Perhaps we are now entering a new age in which the search for truth comes through *fictional* imperatives, such as the one diversely illustrated by Morgantini and Orel in the next pages. Hyperrealistic in its imitation of nature or what is perceived as nature, design has always been made out of unnatural acts.[18]

FROM FUNCTION TO FICTION: QUESTIONING THE INDUSTRIAL OBJECT

What happens to the functionalist orientation of industrial design in this new context? More and more entities of the immaterial society are extrafunctional, multifunctional, function without utility. One basic idea or process can receive many varied treatments.

Design as industrial design in a postindustrial age:[19] What are the conditions, the possibilities, the challenges?[20] Is it possible to talk of informational design? The era of high technology requires an almost indispensable relation to the new machines. As Ezio Manzini has recently written, all high-technology objects will continue to have a surface, a *skin*, which will continue to be charged with emotional and symbolic tensions. Very few people have explored the cultural implications of new material/immaterial objects. "Furthermore," says Manzini, "in intelligent objects we see the results and not the cause. The physical components have become invisible; the mechanism that makes the object work is a minute electronic hearth. The object is seen solely for its performance. It is no longer in front of me, for me to interpret. The relationship changes from asymmetrical to symmetrical and hence specular: I make a move and it makes another."[21] If the main characteristics of the new

Contrasting Features of Industrial and Postindustrial Design

Industrial Design	*Postindustrial Design*
Products are:	Products are:
Specialized	Generalized
Single-purpose	Multipurpose
Short-lived	Long-lived
Replaceable	Repairable
Mass-produced	Short-run
Standardized	Customized
Optimum	Satisfactory
Process is:	Process is:
Autocratic	Democratic
Internalized	Externalized
Exclusive	Inclusive
Intensive	Extensive
Rigid	Relaxed
Designers are:	Designers are:
Creative	Collaborative
Individual	Anonymous
Professional	Participatory

technologies are their pervasiveness and their adaptability, then office work becomes the dominant structural form of work in postindustrial societies. Fast technological change from human- or machine-mediated to computer-mediated work leads to changes in how workers adapt to their particular work environment. Concomitant with these technological shifts are societal and cultural transformations from industrial to postindustrial societies, reflecting changes from a society based on manufacturing and producing material goods to a society based on a service economy (immaterial goods). Under these new conditions, design immediately becomes a more complex and multidisciplinary operation than in the recent past. What is, in fact, to be designed? Chairs, screen, or keyboards, as traditional ergonomics continues to do? Spaces and buildings, as architects have been doing increasingly in the last ten years, proclaiming through the postmodernist vulgate their farewell to the remnants of our industrial societies?

DESIGN/DA-SEIN/PROJECT

Beside the question of *science* is there a possibility for autonomous design? What would that be? Is it the autonomous expression of the project, that dual, ambiguous process that was defined by Leon Battista Alberti,[22] or should we follow Martin Heidegger when he writes of a growing separation between project and reality as an essential condition of our technological modernity? Conversely, Lyotard describes our postmodern condition as the age of immaterials, and Baudrillard talks about the era of the fake and simulacra. It seems that there is a convergence that allows us to define, as a provocation, the sphere of design as the sphere of fake. Industrial design is not a problem-oriented discipline as before, or at least not as evident as it was before. To put things bluntly, we are moving from good form and function to a culture of immateriality and multiple representations. This has been aptly termed as the downfall of the logical principle of coherence, characterized by the simultaneous presence of opposed, contradictory phenomena, such as "the growing automatization of human tasks in relation to machinery and equipment with which they are performed and, at the same time, research in design (almost programmed) for lyrical values (. . .), the breakdown of *stylistic unity* in the neobaroque/postmodern design products, where the various parts are synchronized on a linguistic wave length different from that of the whole."[23]

FORM S(T)IMULATES FUNCTION

If we look at what is going on in the present debate between the tenants of the modernist stance and the neobaroques, we should ask ourselves what becomes of the notion of function: extrafunctional, multifunctional, functions without unity?

Flexibility versus complexity, or should we say versus perplexity: few ideas, many variations. Why is there no theoretical debate in the field of design, similar to the one observed in such diverse fields as architecture or philosophy?[24] Maybe there is no such thing as *autonomous design* and, even less, a *science of design*. Observe what is going on in the debate mentioned above between the tenants of the modernist vulgate[25] and the new romantics or neobaroques. It is an ideological night battle, with invisible and impossible enemies.

DOES ALL THAT PRECEDES AFFECT DESIGNERS? AND, IF YES, HOW?

Before going further into reading the individual essays, we should also consider the professional outlook for the future of designers. Design is changing, and the role of designers is also changing, almost too fast. To clarify some of the designer's dilemma in an age of immaterial complexity, let us use Emilio Ambasz's classification distinguishing three types of designer[26] and follow their evolution in these troubled times. The most exposed type is clearly the third one, the contesting designer. This oscillates between various types of commitment that for the sake of clarity we can summarize as follows. On one side, an extreme position refuses even the category of object of design and reaches for a moratorium, for an antiobject position; on the other side we have the positions of those who view the user-designer-object relation as a unique process, a varying combination with changing patterns of relationship. One can easily suspect that the idea of *new product environment*[27] bears some intellectual and *political*[28] closeness to this position. How realistic and optimistic the position is (at least in the view of some disillusioned practitioners) is evident when we read the *fiction of design*[29] or even, on a less personal base/bias, the dilemmas of technology and participation.[30]

Where can we find the intellectual basis for the new knowledge? It is not, in fact, because we address the issue of the end of the millennium that we adopt immediately a millenarian position. After all, perhaps bareness is the involuntary and paradoxical position when objects, or even entire social processes, become evanescent: from the handling of money to the digitalization of mankind. The evanescence of objects introduces, almost forces, us into a new age for trust, based on blind faith. What else is left when the most basic of our senses are not capable of recognizing what surrounds us?[31]

NEW ROADS, OR ROADS NOT TAKEN BEFORE?

In the last ten years a growing quantity of writings on postindustrial design has almost always presented at the end a series of ritual interrogations about paradigmatic shifts, new roles for design, and, alas, very little else. Who is going to decide the content, mode of access, and operation of programs and procedures that are the basis for so many organizations? What kind of competencies are going to be required to design working spaces, working processes, and products that are eminently cognitive and immaterial? The question calls for a new science of design, resulting from interdisciplinary research, with contributions stemming from many different disciplines,[32] to manage the architecture of complexity that is the most important and vulnerable element of a society dominated by computer-mediated work. Nigel Cross wrote, "Industrial design matured 50 years ago into the Modern Movement, whose standards and ethics we have now become used to accepting. It may be another 50 years before post industrial design reaches a similar maturity, but its seeds are now being sown."[33] Do we really have (the time) to wait another fifty years?

Notes

[1]The almost too obvious reference, and homage, is to Siegfried Giedion's *Mechanization Takes Command* (New York: Oxford University Press, 1948), never quoted and used enough. This outstanding survey of the production of the instruments of material culture probably does not apply any longer in the framework of our dematerialized contemporary society. But no one seems to be able to reproduce such titanic intellectual endeavor: Is it still possible? For partial attempts to answer, see Reyner Banham's *Theory and Design in the First Machine Age* (London: Architectural Press, 1960). More recently, and in a different direction, Jean-François Lyotard tried to sum up the fragmented knowledge of our epoch in his difficult and controversial *Les immatériaux* (Paris: Editions du Centre Pompidou, 1985). A strong dissenting, challenging position is represented by Tomás Maldonado, *Il futuro della modernità* (Milan: Feltrinelli, 1987). A more complete set of references is in the Bibliographical Essay at the end of this book and in the next footnote.

[2]Some of the reflections addressed in these pages have benefited from comments and discussions at a conference sponsored by the Center for Interdisciplinary Research in the Arts (CIRA) at Northwestern University on "Design, Technology and the Future of Post Industrial Societies". A theoretical debate on the nature and the role of design in a period in which we move from industrial to postindustrial societies has already started, notwithstanding Morgantini's pungent provocation/definition of "desert of design." One could refer, for instance, to Barrie Evans, James A. Powell, Reg Talbot, eds., *Changing Design* (New York: Wiley, 1982); Patrick Whitney with Cheryl Kent, eds., *Design in the Information Environment. How Computing Is Changing the Problems, Processes and Theories of Design* (New York: Knopf, 1985); the special issue of *MODO* (English edition, May 1986) devoted to the "Changing Workplace"; the special issue devoted to "Design for the end of the century," *Cahiers du CCI*, 2 (Paris: Centre de Création Industrielle, November 1986). A growing number of international conferences are devoted to the integration of ergonomics, industrial design, and manufacturing in the "evolution of the electronic workplace." The proceedings have been published in *Behavior & Information Technology*, 3, no. 4 (October–December 1984), "Ergonomics and Design in the Electronic Office"; and 6, no. 3 (July–September 1987), "The Evolution of the Electronic Workplace." But this new "consciousness" is still very weak and not very influential. See, for instance Victor Margolin, ed., *Design Discourse: History Theory Criticism*, (Cambridge, Mass.: MIT Press, 1984; Chicago: University of Chicago Press, 1989). See also Andrea Branzi, *The Hot House*, especially Chapter XVIII, "Toward a New Design," and the most recent activities of Domus Academy, among others their "First Contributions to the Third Millennium." If the reader finds too many references to Italian authors, this should be ascribed especially to the fact that debate about design in Italy already involves very large segments of the intellectual and academic community. For an idea of the different level of discussion there and in Germany, for instance, one could refer to Gert Selle, *Design-Geschichte in Deutschland. Produktkultur als Entwurf und Erfahrung* (Köln: DuMont Buchverlag, 1987). The same could apply to most of the American productions quoted above.

[3]I refer to Queneau's *Mille milliards de poèmes* (Paris: Gallimard, 1966). Paradoxically enough, literature has never been too preoccupied by the "materials of life" such as miserable objects of all sorts. There is much research to do on the intellectual sources for the new (design) age, ranging from the already quoted author to very idiosyncratic thinkers and writers such as Georges Perec and Italo Calvino, who already in the early 1960s were forerunners of today's discussions.

[4]Just for the sake of clarity, we indicate hereafter the "problematic areas" that the contributors primarily addressed: the "paradigmatic shift" from modern to postmodern, addressed mostly by Ferraris, Moles, and Morgantini; the question of "participatory design," addressed mostly by Margolin and Iadin, with a special emphasis on knowledge, power, and education; the "nature of design: science, technology or art?" addressed in the chapters by Simon and Wiener and, in part, by Moles; the new esthetic movements

and the consequences for design, addressed by both Forest and Simeoni, with a strong emphasis on the cognition frontier; and the more general societal issues, addressed mostly by Scardigli, Kintzelé, and Maffesoli.

[5]If this societal shift is as real and deep as we assume, then here lies the major justification for the variety of the contributors, who are selected from very diverse ranges of fields, countries, academic surroundings, and intellectual environment.

"Diversity" then becomes one of the key indications of the changing context in which "design" will have to operate in the not-so-remote future. On this topic see the rather acritical perspective of Katherine J. McCoy, "The Landscape of the Future: Speculations on the Post Industrial Age," *Innovation* (Fall 1986), pp. 2–5.

[6]See Alessandro Mendini, "Design pittorico," *Alfabeta*, supplement *Homo faber* (Design ricerca), 88 (September 1986), 15; and Enzo Frateili, Alda Adriano, Chiara Comuzio, Luciano Salio, Annamaria Talanti, "Dove va il design?", *DOMUS* (May 1983), p. 645. See also Enzo Frateili, *Il disegno industriale italiano (1928–1981). Quasi una storia ideologica* (Turin: CELID, 1983).

[7]See further the chapters by Herbert Simon and Norbert Wiener. These show the complex intellectual roots and the *societal* implications in the study of the interrelations between design, technology, and society. They represent almost opposite sides of a large spectrum of theoretical positions. Both Simon and Wiener have played key roles in this debate since its inception: Simon as one of the founding fathers of AI, artificial intelligence, and Wiener as the inventor of cybernetics. See also Ron Levy, "Science, Technology and Design," *Design Studies*, 6, no. 2 (April 1985), 66–72.

[8]The expression was first used in a systematic way by Herbert A. Simon, in his 1968 Karl Taylor Compton Lectures, delivered at Massachusetts Institute of Technology, published as *The Sciences of the Artificial.* (2nd ed.) (Cambridge, Mass.: MIT Press, 1969 and 1981).

[9]To paraphrase the title of the essay by Nigel Cross, John Naughton, and David Walker in *Design Studies*, 2, no. 4 (October 1981), p. 195–201. This article contains an interesting discussion of the topic, even though the authors seek to establish design as a technology rather than a science, drawing on the idea that both activities involve types of knowledge other than just the purely scientific method.

[10]In very fine print, the list of his publications runs for more than thirty pages and covers only the period from 1937 to 1984. Herbert Simon, a real Renaissance man, is Richard King Mellon Professor of Computer Science and Psychology at Carnegie-Mellon University. His research interests span the disciplines of computer science, psychology, organization, and business administration, as well as economics. In 1978, Professor Simon was awarded the Nobel Prize for his "pioneering work on decision-making processes in economic organization." Among his most recent publications are *Administrative Behavior* (3rd ed) (New York: The Free Press, 1976); *The New Science of Management Decision* (enlarged ed.) (Englewood Cliffs, N.J.: Prentice-Hall, 1977; *Models of Discovery* (Boston: Reidel, 1977); *Models of Thought* (New Haven: Yale University Press, 1979); *Models of Bounded Rationality* (Cambridge, Mass.: MIT Press, 1982), 2 volumes; and *Reason in Human Affairs* (Stanford, Calif.: Stanford University Press, 1983).

[11]Simon is paradoxical here for immediately stressing the direct relationship between design and artificiality, a topic that is even more alive today than a quarter of a century ago, as the chapters in this book show.

[12]I have addressed some of these issues in "Conséquences organisationnelles des nouvelles technologies de l'automation," *Culture Technique*, 17 (March 1987), 180–85; "The Social Design of Office Automation," *Design Issues*, III, no. 2 (1987), 73–85, and more recently in *L'intelligenza dell' automazione* (Milan: Franco Angeli, 1991).

[13]See Renzo Zorzi, "Forma come comunicazione," *Alfabeta*, p. 3.

[14]See Ferraris, Morgantini, Moles, and Maffesoli, all "attacking" the modern project from different angles.

[15]See Kintzelé and Simeoni.

[16]See Scardigli and, on the more ironic mode, Iadin.

[17]Arnold Toynbee, *Il racconto dell'uomo* (Milan: Garzanti, 1970), p. 12.

[18]Just reread the illuminating pages by Simon and Wiener to see how far apart the same words can lead. Are we supposed to use what is *natural* in the *natural languages* in artificial intelligence? And why is natural language to be considered natural after all? In the field of design, there are too many automatic assumptions that might profit from being discussed at length. In that respect, and in relation to the question of languages, the contribution of Simeoni is central for addressing a whole new family of questions facing the design field.

[19]See Ezio Manzini, *The Materials of Invention* (Cambridge, Mass.: MIT Press, 1988), 256 pages.

[20]See Patrick Whitney and Cheryl Kent, eds., *Design in the Information Environment. How Computing Is Changing the Problems, Processes and Theories of Design* (New York: Knopf, 1985), 175 pages.

[21]Manzini, *The Materials of Invention.*

[22]It is difficult to render certain Italian terms, the most notable being *progetto*, a word that can mean many others, such as vision, engineering, and so on: See Manlio Brusatin, "Disegno/progetto" in Ruggiero Romano, ed., *Enciclopedia*, 4 (Turin: Einaudi, 1978), 1098–1152.

[23]Frateili et al., "Dove va il design?"

[24]Is it only because "as a form of cultural expression, the design of an industrial product is at a disadvantage with respect to architecture? The latter has thousands of years of history to its name, whereas industrial design has only a hundred years. Consequently, its analytical, critical, and historiographical tools are far less developed and refined than architecture's." Renzo Zorzi, "Ma che cosa comunica il design?", *DOMUS*, 686 (September 1987), 18–28.

On the "coming crisis of architectural theories," see Marco Diani and Catherine Ingraham, eds., *Restructuring Architectural Theory* (Evanston, IL: Northwestern University Press, 1989). The crisis of the civilization of the objects has invested in a particularly violent way the field of industrial design. See also the entire special issue, "Design," *ARCA*, (September 1987), in particular the articles by Gillo Dorfles and Maurizio Morgantini, and the latter's contribution in this book.

[25]A kind of revised version of the combination of ergonomy, human factors, users' needs, and *gute form*. See the various critical approaches in Margolin and Iadin.

[26]The three ideal types are distinguished by Emilio Ambasz: the confirmist, who does not question the sociocultural context of his or her work and focuses mostly on the esthetic qualities of single objects; the reformist, who has profound concerns for the role of designers in society but lives, institutionally and culturally, a profound conflict between their concern and the social and organizational practice of the trade; and last, the contesting one, convinced that there can be no renovation of design until deep and wider structural changes occur. They redesign conventional objects with new, ironic, self-deprecating sociocultural and esthetic reference. See Emilio Ambasz, ed., *The New Domestic Landscape. Achievements and Problems of Italian Design* (New York: Museum of Modern Art, 1972). See also, for a "paradigmatic shift," Branzi, *The Hot House*, especially Chapter XVIII, "Toward a New Design," and Abraham Moles, *Psychologie du kitsch. L'art du bonheur* (Paris: Denoël/Gonthier, 1971), where a form/good taste, a reflection on "taste" as a social expression and cultural artifact, has still to come in the design field.

[27]See Margolin's essay in this issue.

[28]One could object to the use of political in this context: After all, is design *political*? I am using the term in the same way in which it is used by Professor Langdon Winner, Rensselaer Polytechnic Institute, when he talks of *political ergonomy*. See his *Autonomous Technology* (Cambridge, Mass.: MIT Press, 1977). I am grateful to Professor Winner for our stimulating conversation on this subject at Stanford University in April 1987. One can only regret that previous commitments kept him from contributing to this book.

[29]See Iadin's essay and Alan Lipman and Howard Harris, "Dystopian esthetics—a refusal from nowhere," *Design Studies*, 8, no. 1 (January 1987), 33–40.

[30]See Iadin and, to some extent, also Margolin and Scardigli.

[31]See Kintzelé and Scardigli and, to some extent, Morgantini and Moles.

[32]A list of these disciplines would normally "include engineering, computer science, industrial psychology and sociology, ergonomics, technology assessment, management science, economics, system science, social and economic history, political science, and probably others." See Howard Rosembrock et al., *New Technology: Society, Employment and Skills* (London: Council for Science and Technology, 1981). Naturally the author concluded that there is no study in which the different disciplines are brought in contact with one another: "Until this is done, the results obtained within a single discipline are likely to be highly misleading" (Rosembrock et al., *New Technology: Society, Employment and Skills*). See also Claudio Ciborra, "The Social Costs of Information Technology and Participation in Systems Design," in Ulrich Briefs, Claudio Ciborra, Leslie Schneider, eds., *Systems Design by the Users, with the Users, for the Users* (Amsterdam: North-Holland, 1983).

[33]"The Coming of Post Industrial Design," *Design Studies*, 2, no. 1 (January 1981), 3–8.

Postmodernism and the Deconstruction of Modernism

Maurizio Ferraris

Translated from the Italian by Anna Taraboletti Segre

Postmodernism has an ancient history. This fact is not postmodernism's only paradox, although it is the best known and certainly the richest in philosophical consequences. Since Jean-François Lyotard's 1979 pamphlet, *La Condition post-moderne*, which initiated the philosophical and European debate on postmodernism, philosophical postmodernism has been characterized (at least by its promoters) as the consequence of a dissolution of modernist projects, rather than as a new age's allergic, polemic reaction to modernism. If prephilosophical postmodernism (the name given in the 1930s to the Spanish literary period from 1905 to 1914), in architecture or literature, is passably ancient, philosophical postmodernism undoubtedly originated much earlier, from the modernity of the Enlightenment and idealism, with their desire to break with the past and go beyond and their inner contradictions. In short, the birth of modernism sanctioned the birth of postmodernism, just as Plato's dialectic marked the beginning of sophistry or its differentiation from and ostracism by philosophy.

Postmodernism, however, as an explicitly debated and thematized philosophical problem is not old. It originated between 1979, the year of Lyotard's pamphlet, and 1980, the year of Jürgen Habermas's conference on modernism. Before these dates, postmodernism was not a philosophical *topos* but belonged to the tradition of "localized" disciplines, such as literature or architecture. By entering the realm of philosophy, the modernism/postmodernism alternative abandoned its localized character and acquired universal significance (although in certain cases, postmodernism calls into question philosophy's claim to universality—another paradox that will be discussed). As a philosophical theme, postmodernism—or better, the modernism/postmodernism alternative—brings together many earlier philosophical themes, such as the Enlightenment, idealism, the technology-metaphysics relationship, and the concepts of progress and secularization. This history of themes explains how such a recent philosophical theme could appear so ancient, even at the moment it asserted itself.

The philosophical debate surrounding the modernism/postmodernism alternative, which, among other things, deals with philosophy's claim to universality, has universal scope. However, the debate takes place within diverse traditions, nearly to the point that this pluralism prevails over the claim to universality of the philosophical tradition and its channels of communication. This problem forms the structure of this study, which suggests four ways postmodernism has been received in four culturally different contexts: French poststructuralism (Lyotard and Jacques Derrida will be discussed), Habermas's version of critical theory, American neopragmatism (with Richard Rorty and Stanley Fish), and the "weak thought" suggested by Gianni Vattimo in Italy and rooted in Martin Heidegger and Hans Georg Gadamer.

LYOTARD. THE CRISIS OF THE "RÉCITS" OF IDEALISM AND THE ENLIGHTENMENT

Before starting with Lyotard, who spurred the debate, a further short premise is necessary. In prephilosophical applications, the definition or category of postmodernism oscillated between two heterogeneous meanings, *historical* and *metahistorical*. In the former, postmodernism represented a positive overcoming of the modernity of the Enlightenment and romanticism. This definition is, however, contradictory, as modernity developed a poignant emphasis toward "overcoming"—the overcoming of previous ages is the specific pathos of modernity—unlike earlier currents of thought. In the metahistorical meaning of the term, postmodernism originated from the dissolution of the values of modernity rooted in the Enlightenment and idealism, including the pathos of evolution and overcoming, the idea of progress and human liberation, the variations on the death-of-the-arts prophecy, and the obsolescence of philosophy. This dual meaning will be useful throughout my survey. In architecture and in the visual arts, the debate has often favored the historical but philosophically untenable meaning of postmodernism. In philosophy, the postmodernists have emphasized the metahistorical definition of the term, whereas modernists have shown a tendency to see philosophical postmodernism as historically self-explanatory, a fact that has led to numerous misunderstandings.

In *The Postmodern Condition,* Lyotard was undoubtedly referring to the metahistorical definition. His analysis did not focus on the positive definition of the category of postmodernity but rather on analyzing the two main discourses (in Lyotard's words, *récits,* "narrative accounts") that presided over the philosophical and political justifications of knowledge in modernity, from the eighteenth century onward. On the one hand, the discourse of the Enlightenment saw knowledge as the vehicle for the liberation of humanity, which left infancy and freed itself from the obscurantism of tradition. On the other hand, the idealist discourse justified knowledge only inasmuch as it could be included speculatively within a doctrine of science (the philosophical one). Knowledge would then be placed in an absolute perspective, deprived of unilateral or instrumental empirical ends.

Following the presuppositions of a metahistorical understanding of postmodernism, Lyotard aimed to demonstrate how the same causes that led to the dissolution of the modern discourse actually were *immanent* in the Enlightenment and idealism. In this, Lyotard places himself in the well-marked path of philosophical discussion on modernity—in that radical philosophical current that starts from the philosophies of the Enlightenment and transcendental idealism and develops their inner contradictions, as well as their potential for self-dissolution. One needs only to think of Friedrich W. Nietzsche and his twentieth-century heirs, such as Heidegger, who based their criticism of the modern metaphysical tradition on the monumental work of a historical rereading of modern and ancient metaphysics. Or one thinks of the opposite end of the same tradition, Max Horkheimer and Theodor Adorno, who criticized dialectics and the Enlightenment in the name of dialectics and the Enlightenment. That is why philosophical postmodernism is so ancient and not as detached from modernism as its critics claims.

Undoubtedly, Lyotard's interpretation of the dissolution of modernity's two *récits* can be traced from the canons of radical philosophy of the nineteenth and twentieth centuries. The only estranging effect is the framework in which the discussion takes place, not philosophical historiography but rather the sociology of knowledge, chosen as a tribute to the American public it initially addressed. As for the *récits* of the Enlightenment, in its struggle against myth and prejudice to dispossess tradition, it denounces not only the myths of the past but also its own: the mythologies of reason that translate to a mythical horror of myth, a systematic effort to deprive tradition of its authority, and a unilateral emphasis on the liberating power of reason. Yet reason could not promise liberation *ideo facto*. The Enlightenment became an instrument in the hands of great statesmen such as Frederick the Great or the Jacobians, who used it to establish a new obscurantism and a more properly modern terror. Above all, there is no guarantee that such a rational and descriptive game might lead peacefully to prescriptive, practical, and liberating consequences. It is precisely this point that Lyotard emphasizes.

As for the *récit* of idealism, with its project for philosophy as a doctrine of science, transcendental idealism assumed that the positive knowledge of the sciences did not have value of its own but only inasmuch as it could be understood in a speculative manner, within the framework of an absolute philosophical knowledge. Taking justification upon itself, transcendental idealism removed responsibility from single disciplines. Exactly at the time when a speculative doctrine of science was found to be unattainable, science as positive knowledge was itself deprived of any imminent justification (because justifications had to be external, according to the idealistic concept of knowledge). In this, Lyotard relied on Nietzsche's perspective (particularly on his 1872 conference *On the Future of Our Educational Institutions*). But in this theme, the fading of absolute knowledge and the justification of science, a common thread leads from Nietzsche to Wilhelm Dilthey and to the search for a hermeneutics to use as the systematic scheme of the sciences of the spirit, to Heidegger's radical and certainly extreme condemnation of a scientific knowledge deprived of speculative value. However, the discourse on delegitimizing transcendental idealism as a scientific doctrine and an encyclopedia of philosophical sciences refers to a process of inner decay. In Lyotard, this discourse is not opposed to modern science (as some critics of postmodernism perceived). And certainly, even in Heidegger, if science does not think, it is because, in general, philosophy has yet to begin thinking.

HABERMAS AND MODERNISM AS AN UNFINISHED PROJECT

A major opponent of Lyotard's theory was Habermas, who, since a famous lecture in 1980, criticized architectural and later philosophical postmodernism, and subsequently clarified his position analytically in numerous works (particularly in a collection of lectures, *The Philosophical Discourse of Modernity*, 1985). Habermas's perspective can be summarized as follows: Postmodernism is not autonomously legitimate but rather is the consequence of delegitimization of modern philosophical discourses. Therefore, it qualifies simply as *a sign of the times*, whose justification must be sought externally in the incompleteness of modern philosophical discourses (or rather in the one philosophical discourse of modernism, which, in Habermas's view, is the Enlightenment). Thus, postmodernism must be considered only as a symptom of the long-time status of the project of the Enlightenment—a project that must, in any case, be completed to prevent the rise of obscurantist and deliberately conservative positions.

Habermas's dismissal of the postmodernists as "signs of the times" is quite different from Gyorgy Lukacs's criticism of the artistic avant garde; it is not based on heterogeneous presuppositions of the analyzed phenomenon but shares the postmodernists' basic diagnosis of the crisis of modernism. The project of the Enlightenment stalled and almost died because of its own aporias, which were brought to light by the radical philosophy inspired by the late nineteenth and early twentieth centuries. In this process, Nietzsche, as well as Horkheimer and Adorno's definition of the dialectic of Enlightenment, played a decisive role. Such a position was already present in Habermas's *Knowledge and Human Interests*. In particular, Nietzsche drew conclusions from the aporias of the idealist project on absolute knowledge and the project of the Enlightenment on liberation through reason. In doing so, however, he did not rely on regression, the mystic-feudal concept, the return to the earth: "His criticism of philosophy of science, his criticism of the predominating ethics are the only evidence of knowledge attained through self-reflection and only through self-reflection."

As noted above, Lyotard's position is no different: The *récits* of idealism and the Enlightenment are weakened by the struggle of reason against prejudice. This perspective explains the recent developments of radical thought. The *Dialectic of Enlightenment* serves to analyze the collusion between the Enlightenment, totalitarian obscurantism, and the new mythologies induced by a mythical horror of myth, as well as Michel Foucault's microphysics of the power/knowledge relationship and Derrida's grammatology as the definition of a spirit that is rational and objective but internally opaque and without foundation. At this point, Habermas and the postmodernists take different paths. According to Adorno, Foucault, and Derrida, it is necessary to pursue *deconstruction*, a criticism to practical reason that weakens the claims of reason and questions the totalitarian eventualities of humanism. This prescription is based on the presupposition that the difference between pure reason and practical reason is extremely vague. It does not suggest alternative utopias, which could become new instruments of power since the revival of the Enlightenment. The mythologies of blood, earth, or technology cannot escape the insidious dialectics of the Enlightenment. In Heidegger's own terms, metaphysics (which is, among other things, the call of reason to power and total organization) can be overcome only by starting afresh with it, but with a fresh start that involves loosening, weakening, and distorting it.

With different nuances, Habermas dismisses all this—deconstruction, specifically determined negation, and the relationship between overcoming and relying—as the same old tune of self-overcoming and suggests again the link between reason and liberation. That is, every kind of reason lends itself to exploitation. But this should not prevent us from becoming interested in liberation as a regulating interest and a practical postulate that can justify philosophy and at the same time protect it from regressive or totalitarian temptations. If being enlightened means to become aware of the limits of the reason of the Enlightenment, to continue the project of reason and human liberation today constitutes the only objective and justification of thought.

In this sense, not all defections from the project of the Enlightenment are equal, and Habermas claims the need to trace a line of demarcation. On one side, the current of radical philosophy remains the most faithful to the philosophical project of modernity (Adorno as its representative, "remains, desperately perseverant, faithful to the procedure of determinate negation, faithful to the idea that there are no remedies to the wounds of the Enlightenment, except for the radicalized Enlightenment itself.") On the other side, Foucault, Derrida, and the poststructuralists, in general, favor instances that are, according to Habermas, heterogeneous with respect to the most genuine project of the Enlightenment. Examples of such instances are an appeal toward the will to power (Foucault and, earlier, Nietzsche) or the irrational demonization of the rationality of the Enlightenment (Derrida and, earlier, Heidegger).

POSTMODERNISM IS THE DECONSTRUCTION OF MODERNISM

Referring to what Habermas defined as the topic of a transcendental *tu quoque*, it would be easy at this point to notice that self-overcoming is here dismissed as an old tune precisely on behalf of the other, historically more ancient tune of liberation. But this reasoning is not good philosophically. Rather, two points that are important to the relationship between critical theory and poststructuralism should be restated.

Liberation and pragmatism. Habermas's proposal to reconstruct the modern project of liberation is not only nostalgic (its most significant trait) but also pragmatic and pragmatistic. To explain further: On the historical, theoretical, and moral levels, the interest in liberation as suggested by Habermas no longer has anything in common with the original project of the Enlightenment.

According to the thinkers of the Enlightenment in Denis Diderot's and Jean-Le Rond d'Alembert's time, the issue of liberation could be logically deduced from a clearly structured situation, formed by universal reason that would free European humanity. (Slaves were not included, whereas a few token Persians were present merely for literary theme.) The issue could also be deduced from a meaningful or self-explanatory and taken-for-granted history, according to which the customs of nations were slowly emerging from barbarism.

Romanticism and idealism were not simply the practical and instinctive reaction of kings, people, and the bourgeoisie against terror but also the philosophical acknowledgment of the unilateral character of perspectives rooted in the Enlightenment. Absolute knowledge, not a science but an encyclopedic doctrine of science, became necessary to free the goals of enlightened emancipation from their unilateral character. From Hegel to the avant garde, romanticism implied not only traditionalism and a return to premodernism but, above all, a hypermodern drive to go beyond the aporias of modernism.

The failure of the project on absolute knowledge was recorded by Nietzsche, Dilthey, and the radical philosophy of the nineteenth and twentieth centuries. Here also, for reasons mentioned in the introduction, recording failure does not mean to regress but to examine thoroughly and at the same time to go beyond the contradictions of modernism. In short, postmodernism is the deconstruction of modernism. The perspective of reconstruction adopted by Habermas shows instead an inversion of this pattern of evolution, a sort of Copernican revolution. The interest in liberation, returned to its historical matrix, no longer constitutes epistemological evidence but rather a moral absolute, prescribing that if you persevere in your pursuit of emancipation, you will be able to find history, reason, and the human species. Once an effect of a little-analyzed situation, the interest in liberation has now become a cause—an efficient cause more than a good or ideal cause capable of directing philosophy. From this comes Habermas's dogmatic condemnation of any deconstructive perspective, which he qualifies inexplicably as being regressive.

Reconstruction and deconstruction. Space is lacking to evaluate the choice between reconstruction and deconstruction of modernism (another way to define postmodernism). Therefore, only a schematic description will be presented. Heidegger's *Destruktion* was not simply a destruction of metaphysics (thus Derrida chose to translate the term as *deconstruction*) but also a historical-critical retracing of metaphysics. (As Heidegger wrote for a 1927 summer course, "Philosophical construction is necessarily destruction, that is, deconstruction, performed through an historical return to tradition, of what has been handed down; this does not in the least imply a denial of the tradition which accuses it of nullity, but rather a positive appropriation of this tradition.") In the same way, a philosophical discourse on the modernism/postmodernism alternative should present itself as a deconstruction of the modernity of the Enlightenment and idealism. A dismantling and different reconstruction, not prejudiced by preexisting patterns, was already the project of transcendental idealism, especially Hegel's: The reconstruction of history of philosophy was leading to a completion that was substantially an overcoming, at least of the perspective of the Enlightenment. Those who determine the limits of reason have already gone beyond them; this is Hegel's famous criticism of Immanuel Kant. Such has also been the project of radical philosophy, from Nietzsche on, never confining itself to the simple exclusion of modernity and reason—as Habermas acknowledged, at least, for Nietzsche and Adorno—but always acting deconstructively.

In his 1962 monograph on Nietzsche, Gilles Deleuze wrote, "It is a serious mistake to think that irrationalism opposes to reason something that is not thought—whether it be the rights of the given, of the heart, of feeling, caprice, or passion. In irrationalism we are concerned only with thought, only with thinking. What is opposed to reason is thought itself; what is opposed to the reasonable being is the thinker himself." The object of this metacriticism of knowledge is to impose limits, which, by the very same act of being imposed, are already surpassed. The same happens for postmodernism when it justifies itself as the deconstruction of modernism. One must accept (and the modernists are the first who should do so) that postmodernism is not an enormous departure from modernism but rather a capillary system of revisions and transformations. Without excluding philosophical tradition and the modernism of the Enlightenment (an exclusion of this kind would only imply blindness toward presuppositions, or an oblivion that Heidegger qualified as the absence of metaphysics) postmodernism subjects modernism and tradition to a micrologic revision. This task was undertaken by Derrida, in particular, and is not too distant from Adorno's Enlightenment dialectic. Between negative dialectics and deconstruction there are no conceptual distinctions, as Habermas maintains, but simply diverse cultural conditions, such as the appreciation of Heidegger, traditionally ostracized in German philosophy but not in French philosophy. The reasons for this ostracism were political rather than speculative.

RICHARD RORTY AND STANLEY FISH. POSTMODERNISM AND PRAGMATISM AS THE EXCLUSION OF PHILOSOPHY

The above discussion can be confirmed perhaps by the examples of Richard Rorty and Stanley Fish's neopragmatism. Habermas is more indulgent toward Rorty than toward poststructuralism, for reasons that are cultural, in my opinion, not philosophical. Rorty is preferable to Derrida because his discourse lacks antihumanistic implications, and this fact alone mitigates the strictness of his conservatism. But upon more careful scrutiny, the projects of Rorty and Fish are the specular opposite of Habermas's discourse, although they all misunderstand the deconstructive issues of postmodernism.

To explain further, Habermas proposes a reconstruction of modernism and philosophy, whereas Rorty and Fish, who declare themselves postmodernists, do not see postmodernism as a complex process of self-overcoming of modernism but a simple exclusion of modernity and its philosophy. Such an exclusion focuses on two key points of modernism in the Enlightenment and idealism: the practical significance of philosophy and its claim to universality. In *Critique* (1983), Rorty intervened in the controversy between Habermas and Lyotard, maintaining that, despite their divergent opinions, the two are linked by an overly binding presupposition: the belief in philosophy's universal destiny, by which decisions made under philosophical aegis would constrain the future of humanity. And what if, adds Rorty, there is agreement on the fact that the outcome of postmodernism is precisely the deemphasis on the universality and practicality of philosophy? What if the conclusion is reached that the discourse on the intellectual's moral and practical engagement is not a symptom of the universal and political call of philosophy but rather a form of self-assurance adopted at times by certain intellectual circles? What if the philosophies of modernity, such as idealism and the Enlightenment, were to be numbered among the inventions and gratifications of the bourgeoisie, together with psychological drama, literary dandyism, and figurative modernism?

With an undoubtedly paradoxical pragmatism that excludes the practicality of philosophy, Rorty invites consideration of philosophy as a genre of writing that, more than any other medium of communication, lacks any privileged relation to the truth. (According to Rorty, the most noticeable characteristic of Derrida's writing, a playful attitude, announces what philosophy will become after the crisis of the *récit* of modernism.) Alternatively, restating a topic Rorty developed in his major work, *Philosophy and the Mirror of Nature* (1979), philosophy could possibly be considered as chatter, one conversation among many, containing truth and relevance but only in a relative and contingent way, as might any other discourse.

Stanley Fish's reasoning in *Is There a Text in This Class?* (1980), basically complements Rorty's, moving from the basic thesis that theory has consequences on theory only. Thus a theory does not influence the universal behavior of humanity (which is governed rather by heterogeneous beliefs). It does not even have any ultimate impact on related theories; only through complex, institutional mediation can a judicial or literary theory be influenced by a philosophical theory. Most of the time, we can even observe a reciprocal impermeability between philosophical readings: The ideas of analytic philosophy do not influence those of continental philosophy and vice versa. To stress such a peculiar condition is perhaps the undesired or half-desired effect of antifoundationalism—the trend leading from a genealogy of morality to hermeneutics, deconstructionism, and pragmatism. By questioning its own presuppositions and concluding that there is neither a causal order of the world nor a theory capable of informing praxis, philosophy performs a self-critique and, even more radically, denies its own claim to universality. The discourse of antifoundationalism is the last universal discourse of philosophy, after which philosophy has to accept its role as a simple academic discipline. The problem is that the discourse that relegates philosophy to particular institutions shares with antifoundationalism and all earlier arguments on the death of philosophy that same characteristic of universality. By starting from a universalistic perspective, Rorty denies the universal ambitions of philosophy. Who can guarantee that philosophy is more than a genre of writing if not somebody who implicitly puts himself in the position of looking at all genres from above? Behind the argument that maintains that parliamentarianism is a historically more significant event than the birth of modern philosophy is an extraordinary similarity to the arguments of old positivism (philosophy is a thing of the past; facts are more important) or of the Marxist *récit* (behind philosophy, literature, arts, and so on there is the bourgeoisie, of which these things are expressions).

Rorty and Fish's argument is not basically immune to the objection made against the exclusions of philosophy, an objection claiming that such exclusions are rendered in the same language of philosophy and thus follow the same historical destiny while decreeing its end. (As for the characterization of philosophy as being particular because it is chatter, the argument seems to demonstrate the contrary of what it asserts: In contrast to chemistry, carpentry, or telecommunications, chatter is the typical universal discourse. To argue that philosophy is chatter does not so much limit philosophy and chatter to cafés or universities as it points out that both share a common propensity toward universality.)

GIANNI VATTIMO AND THE "WEAK THOUGHT"

Considering postmodernism as mere reconstruction of modernism (as basically Habermas had it) is not the issue, not as an even more simplistic exclusion of modernity (as Rorty and Fish had it). The basic aporia of these two specular positions favors the view of postmodernism as the deconstruction of modernism and probably supports the "weak thought" suggested by Gianni Vattimo in Italy, which shares, with deconstruction, common roots in Nietzsche and Heidegger. A few basic points of postmodernism as seen in the perspective of the "weak thought" follow:

The impossibility of overcoming modernism. As with the deconstructionists, Vattimo denies that the detachment from modernity is a critical overcoming, as this would be a basically modern perspective. (This is the historical and thus contradictory view of postmodernism, as mentioned at the outset.) As Vattimo writes in *La fine della modernità* (1985), "If modernity defines itself as the age of overcoming, of the novelty which ages and is replaced immediately with a newer novelty—within an unstoppable movement which discourages creativity at the same time demanding and imposing it as the only form of life—if this is what happens, one will not be able to escape Modernity by *overcoming* it."

The impossibility of excluding philosophy. In fact, such an objection is valid not only for the nonphilosophical attempts to conceive historical postmodernism as a critical overcoming of modernism. It also is true for the pragmatic proposal to exclude philosophy, precisely because, contrary to Rorty and Fish's beliefs, the tradition of philosophical modernism does not consist of a doctrinary corpus, easy to handle and consign to a campus. Modernism, as well as philosophy in general, unavoidably permeates the realm of life, beliefs, and justifications. To exclude philosophy by transforming it into literature would not be a postphilosophical gesture but rather the *philosophical* decision of an unhappy—or a self-denying—positivist.

Verwindung/überwindung. With respect to the weak thought, the point is also a matter of conceiving postmodernism as the deconstruction of modernism, as an overcoming of the tradition of modernism. This does not intend to exhibit new contents and values for a new epoch but rather to retrace the myths and limits of modernity—as an abandonment of the specifically modern pathos for the new one. This model is the same as the one adopted by Heidegger for the problem of overcoming (*überwindung*) metaphysics. A similar *überwindung* cannot be proposed as a critical overcoming or a simple exclusion of metaphysics, which would reappear, denying itself, precisely where it is supposed to be overcome. As Heidegger suggests, *überwindung* should be considered as *verwindung*, not as overcoming but rather as relying upon the metaphysical tradition and the tradition of modernism above all, in such a way as to imply distortion, unbinding, and weakening of the pathos of metaphysics and modernism, undoubtedly beginning with the pathos of novelty.

A few final comments on the problems of postmodernism as the deconstruction of modernism follow.

Within limits that are objectively difficult to define, in philosophy as well as in any other field, modernism and postmodernism play a *rhetorical* role rather than a historical or philosophical one. This is the same phenomenon that occurred, for instance, in the dialectics/sophistry contraposition (in Plato's time and throughout the history of philosophy), or in the rationalism/irrationalism alternative (in the twentieth century). Each pair of terms includes a good part (dialectics, rationality, and, in my opinion, modernism) and a bad part (sophistry, irrationality, and, in my opinion, postmodernism). However, in each pair, the principle that leads the good term to condemn the bad one does not seem to depend exclusively on the ostracizing term but also on the term being ostracized. To explain further, the discursive arguments Plato used to condemn sophistry in the name of philosophical dialectics originated from dialectics as well as sophistry. (This is a well-known, almost banal fact in platonic exegesis.) Moreover, sophistry plays a supplemental role in the constitution of dialectics, as the philosophical art of true and live discourses cannot exist without the other art of false and dead discourses. Sophistry plays a decisive role as an adverse foundation. Likewise, the contraposition of rationalism and irrationalism is not rational but is a creation of rationalism itself, which finds in it its own justification and value. And what rationalists call irrationalism never actually defines or qualifies itself as irrationalism. (Are there philosophers, or anybody for that matter, who would seriously consider introducing themselves as irrationalists? Things are certainly more complex.)

Thus, in the introduction, the alternative of modernism and postmodernism was compared to that of dialectics and sophistry and, later, of rationality and irrationality. Modernism cannot be thought of without thinking of postmodernism, or postmodernism without modernism; otherwise, their essence—an essence that does not reside in one of them only but in their alternative and, particularly, in their reciprocal parasitism—would be missed. This parasitism can be defined as the *deconstruction of modernism*, or, as Vattimo put it, as *verwindung*.

On the one hand, postmodernism is certainly a parasite of modernism. As the postmodernists admit and the modernists deprecate, postmodernism is undoubtedly a citation of modernism, and, thus, postmodernism is not as young as might be inferred. On the other hand, modernism is not less parasitic toward postmodernism. This apparent paradox can be easily explained. During the Enlightenment, the renewing pathos of modernism was loosed upon archaism, myth, theories of divine right, obscurantism, and so on. In the justification of modernism, archaism played the important role of an adverse foundation. However, already in the age of romanticism, the past no longer appeared primarily as ancient regime but rather as Jacobian modernity. In time, the past as archaism vanished even from the Third World, which is now completely industrialized, if not less poor than before. So at present, postmodernism plays a major role in the self-proclamation of modernism, which finds its value and justification by ostracizing postmodernism. Precisely for this reason, modernism attacks postmodernism, defining it as *neoconservatism*.

I believe, therefore, that modernists should be very grateful to postmodernists, because their ostracism of postmodernism now constitutes the only *point* of the discourse on the reconstruction of modernism. What could Habermas write without Derrida? How topical would his program for the reconstruction of modernism be?

REFERENCES

Deleuze, Gilles, trans. Hugh Tomlinson, *Nietzsche and Philosophy*. New York: Columbia University Press, 1983 (Paris: PUF, 1962).

Fish, Stanley, *Is There a Text in This Class? The Authority of Interpretive Communities*. Cambridge, Mass.: Harvard University Press, 1980.

Habermas, Jürgen, trans. Jeremy J. Shapiro, *Knowledge and Human Interests*. Boston: Beacon Press, 1971 (Frankfurt am Main: Suhrkamp, 1968).

Habermas, Jürgen, trans. Frederick G. Lawrence, *The Philosophical Discourse of Modernity: Twelve Lectures*. Cambridge, Mass. MIT Press, 1987 (Frankfurt am Main: Suhrkamp, 1985).

Lyotard, Jean-François, trans. Geoff Bennington and Brian Massumi, *The Postmodern Condition: A Report on Knowledge*. Minneapolis: University of Minnesota Press, 1984 (Paris: Editions de Minuit, 1979).

Rorty, Richard, *Philosophy and the Mirror of Nature*. Princeton, N.J.: Princeton University Press, 1979.

Vattimo, Gianni, trans. Jon R. Snyder, *The End of Modernity: Nihilism and Hermeneutics in Postmodern Culture*. Baltimore, Md.: Johns Hopkins University Press, 1989 (Milan: Garzanti, 1985).

Design and Immateriality: What of It in a Postindustrial Society?

Abraham A. Moles

Translated from the French by David W. Jacobus

An immaterial culture is emerging. It exists only because a heavily material base supports it and makes it possible. It is from the very outset a phenomenon—indeed, an epiphenomenon—resulting from technology. The future of design, then, for an artificial reality depends on the design of the hardware and specialized techniques. These are the fundamental constituents of an artificial reality that contribute to the creation of what one could call *imago*—generalized images, not necessarily confined to a visual mode. Thus, a postindustrial society is a superindustrialized society, or one which has pushed to extremes the consequences of its industrialization.

It is true that we are surrounded by so-called electric phantoms, to use a phrase of Villiers de l'Isle Adam, which more and more are invading both our work and recreational environments. One of the problems posed to the human spirit is its capacity to exercise control over reality, while adjusting to the blurring of barriers between reality and images, or between real objects and their appearances. As we enter the age of *telepresence*, we seek to establish an equivalence between "actual presence" and "vicarial presence." This vicarial presence is destroying the organizing principle upon which our society has, until now, been constructed. We have called this principle the *law of proximity*: What is close is more important, true, or concrete than what is far away, smaller, and more difficult to access (all other factors being equal).

We are aspiring, henceforth, to a way of life in which the distance between us and objects is becoming irrelevant to our realm of consciousness. In this respect, telepresence also signifies a feeling of equidistance of everyone from everyone else, and from each of us to any world event. At the same time, we live in an age of communicational opulence. We now have at our disposal more sources of communication and interaction than we will ever be able to make use of in our relatively short lifetime. This is the age of a social system of networks, decorated with the futuristic name of the "Information Society." Henceforth, the bulk of our effort will be spent more for manipulating information than for manipulating objects, which are now no more than products of far-out robots, controlled by abtruse programs, inaccessible models, and a ubiquitous creator.

Nevertheless, these new living conditions, which are also cultural conditions (and thus what one used to call the conditions for art), can subsist only on a spectacular hardware or material base. This underlying structure is spectacular, most evidently, because of the astonishing omnipresence and sheer size of the machines and the specialized manufacturing plants making cars, video equipment, audio components, and so on. While generally unaware of the highly specialized, technical origin of such objects, the layman has to live with this material accumulation. But sheer size is not the most significant aspect of this material base. Rather, it is especially spectacular because of the complexity that stands as today's paradigm for technological advancement. (A 3-cm microprocessor comprises more "things" than an automobile—more components, more functions, more connections, more relays, and, conceivably, more raw intelligence.)

Yet the structural complexity of today's hardware, which can be quantitatively measured, finds itself subject to one of the general laws of the universe, the law of entropy, or, simply put, the tendency of all conjunctions of things toward disorder. This irrepressible agitation of the physical world counteracts and eventually destroys the order imposed by the hardware creator. The consequences are overloads, short circuits, and equipment failure.

A MAINTENANCE SOCIETY REPLACES A PERFORMANCE SOCIETY

While transmitting and receiving messages and images constituting an artificial reality, man encounters more frequently (like the *diabolus in machina* ("devil in the machine") of the religious era) the bug, the hitch, and the malfunction, largely random in nature, which erode the fundamental virtue that any mechanical aid or appliance must possess: reliability. All other factors being equal, the more our world is complex, in reality or appearance, the more it is fragile, and the more it is subject to mishaps. As a result, the hardware builder must adopt particular strategies or plans of action designed to control the demon of disorder. Such action gives rise to a *maintenance mentality*, which perfuses the world of support groups and technicians, for making "immaterials."

The creators of the new immaterial culture have devoted most of their effort to the permanent reproduction of what, for them, was once practically inconceivable: the transmission of sounds and speech over distance and the storage of data in memory banks. All this, which a century ago was hailed as "the miracle of communication," has become commonplace in our lives. From a social perspective, the inventors and the builders seem to have won the battle waged on the continual insurrection of nature against the complex pattern of circuits and transistors. They have learned how to incorporate a reasonable degree of reliability into an uncertain miracle and have mastered the multiple and diverse causes of disorder through deduction and quantitative reasoning. The digital compact disk or the satellite numerical repeater are two noteworthy examples. *The immaterial civilization must be reliable*; otherwise it could have no social impact and would be the subject of little or no debate today.

But the minimal reliability established by the industrial complex still does not inspire in the individual sufficient confidence to participate in an immaterial culture. He or she is reluctant to depend completely on telerepresentations being confronted by a heterogeneous array of equipment that originated from various sources, was assembled at random, and offers uncertain compatibility. The situation of the individual, then, is different from that of the society at large, and consequently the meaning of the term *reliability* becomes more strict according to the scope of the human or material circumstances to which the word is applied. To live and experience a culture is to interact with innumerable, disparate forms of that culture, as well as to find oneself presented with a myriad of possibilities for successful—and unsuccessful—operation. The demands that the individual presents to society, therefore, are more important than the demands of society on its subjects, which is the situation we are faced with today.

The task of the designer is, precisely, to ensure reliability by mastering the factors that jeopardize it. Therefore, the first field of investigation of design for an immaterial culture is to furnish largely material assurance of the *universal reliability* of the systems that makes it possible.

DESIGN BY A MODEL OR IMMATERIALISM OF DESIGN?

A second direction of inquiry into the relationship between design and the immaterial culture is to investigate the impact of the stratagem on the designer's *task* in a computerized society. We used to say that the designer, an environmental engineer, was responsible for each of the systems he installed, as well as for their integration into the particular life-style of each person. Each of these systems required a material task of design, done in a workshop that, until now, was hardly susceptible to automation. All this is changing, however, with the advent of computer processing and computer-aided manufacturing, which are modifying the phenomenological *nature of the design process* (Larroche, Tuchny).

It is useless to ask if design is an artistic activity or a scientific one; the etymology of the word *technology*, or *techne* (art), provides a sufficient response. With respect to doing, the confusing etymology of the word *engineer* (the genius who drives the engine or machine) applies also.

The designer creates the environment of others. Until now the designer's vocation has been both conceptual and concrete. The workshop has been a place to perceive and build models destined to be copied for mass production. The question now, however, is how this vocation has been modified, for better or for worse, by the inexorable development of the immaterial culture. The design activity itself is changing because the designer's tools are becoming immaterial, as are the lives of those to whom the products are marketed. Design by holographic models may still be far in the future, but the concrete activity of design even now partakes substantially of immaterial techniques, or usage of artificial representations, images, and diagrams composed by image-generating machines.

Affecting more than workshop activity alone, the trend toward immaterialism includes all projectional conception in a concrete model, a process that used to depend on a situation of permanent interaction between conception and construction. The dialectic game between the abstract (the idea, the mental vision) and the concrete (the struggle with the material and disparate tools and appliances) is giving way to work done essentially with computer-integrated manufacturing at a computer desk. The designer's task now consists of construction based on three precepts: (1) the work order, (2) the rules of an exploratory program, and (3) articulation in a field of liberty defined by certain parameters—often of ergonomic origin and eventually of creative fantasy. This suggests to the designer, regarded as the master of the work, an ensemble of more or less pertinent variations on which he will have to exercise his critical abilities, themselves controlled by functional optimization principles in the broadest sense. The designer, who assumes a more and more important social role in a society where power is evanescent, must confront the new fundamental idea of "initial form plus variations," in order to realize an often monotonous variation of possible forms with respect to given parameters. This is what we have called "variational creativity."

We are passing from a time of hands-on creation of a model to one of an initial form plus a field of variations, which stem from any already existing object, whether traditional or modern. Thus, from this established model a whole series of new models is determined, with the intention of forever satisfying the avidity of the consumer market. This process is done using the most refined and the most abstract creative techniques, in order to express them in programs, through creative methods (Alexander).

Not only the designer, but also the manufacturer of forms and models faces this situation, but with a different perspective. The manufacturer confronts a kind of creation in immateriality, which encompasses the definition of the field of possibilities and its methodical exploration, which, in turn, generates auxiliary criteria for judgment. These criteria are then reapplied in order to furnish yet another new field of possibilities. Henceforth, one expects no longer to find drafting tables, sculptor's tools, or carpenter's chisels in the design room. They are being phased out by drawing and image-creating machines that yield computer graphics, so that the material objects themselves, as products of these images or of audio and visual simulations, are, at a distance, mere products of the imagination, which seem more *credible* than *real.*

From 1850 to 1950, industrialization was characterized by the predominance of a system of drafted plans and diagrams, which were essential to the materialization of ideas and which caused a proliferation of design patents. DeForge, in a noteworthy book, illustrated well this "kingdom of drawings" or technical diagrams, which were the rule in nineteenth-century workshops.

For several decades the complexity and precision of drawings have rendered them amenable to both the computer's memory and the drafting table. Now, however, the draftsman and the modelist are disappearing from the finishing laboratory or the graphic workshop itself; the computer is supplanting them with its ability to generate, on demand, any view, profile, or cross-section of any part or whole of the factory or machine in question. Furthermore, the implicit and intuitive relationship itself of a small part to the whole, or of a particular machine to the whole factory, is vanishing. In many complex systems, the global design breaks loose, becomes accessory, indeed superfluous, because the whole changes with each moment of its construction. Certain parts are replaced by others and modifications are made during construction, even during the useful life of the product itself, because the conditions determining the integration of smaller parts into the whole are automatically taken into account by the computer. In short, the scale model is losing its significance; it serves only as a rough estimate or guide to the builder. It is only a palpable illustration of the concept involved and, most important, is no longer necessary. One could say that neither the designer, the engineer, nor the architect knows the system in all its exact detail, though any one of these persons could simply ask the computer for any piece of information concerning the function of the object under construction. Thus, we are arriving at a kind of conceptual immateriality of complex systems in the wake of those who create them in fragments.

THE DESIGNER AT THE CENTER OF COMBINATIONS OF ESTHETIC EFFECTS

Finally, it seems appropriate to consider the relationship of design with a fundamentally artificial reality. This is another systematic exploration of the field of possibilities and will be one new sensorial combination, generated from a "new art," defined by the sensory parameters that it manipulates and the new esthetic arrangement it proposes. If, in the foreseeable future, art remains (for those unperfused to or allergic to the idea of chance) a programmed sensualization of the environment, the designer, as an environmental engineer, finds himself endowed with considerable ability to manipulate the new artistic matter, a step that transforms the designer himself into a neoartist. Or could the designer be called a meta-artist, the potential maker of new "art"? In this respect, our senses of the close range of those affecting us by contact (touch, smell, sensitivity to temperature, vibration, and balance) remain a relatively unexplored area of human sensoriality. We lack the means to evaluate objectively certain aspects of the real, which we could call *transduction* (to transform messages from one medium to another) or *interfacing* (to set up a partition of illusions for the projection of tele-images for example: a screen, a tactile sensor, a sonorous background, a simulated landscape, or a virtual actor). The immense technological structure proposed by the postindustrial society seems precisely to have to fill quickly this gap. It is becoming the function of design to examine this new field of "programmed sensualizations" (what one used to call a "work of art" and what one could call from now on a "scene of esthetic action"). What would become, for example, of an electronic tactile detector combined with a Minitel or Compuserve system? In a purely philosophical sense, would the result be a by-product of the immaterial culture, or would it be a new event in the sphere of esthetic valorization?

THE POSITIONS OF DESIGN DURING THE AGE OF THE INVASION OF IMMATERIALITY

Any immaterial civilization will be heavily materialized because its immaterial products are necessarily linked to the mechanical infrastructure that generates, stabilizes, and governs them. By misinterpretation, a recent exposition celebrated not an "immaterial world" but a form of the binomial object-images, forms, and support. This is another attitude of man vis-à-vis an artificial world—a world created by him as the only source of his reactions. Does such an attitude imply a danger of technological narcissism, at the moment when nature becomes inaccessible or restricted to parks with limited access?

Thus, every symphony has its compact disk; every audio experience its loudspeaker; every visual image its camera and video disk. Behind every outward image or symbol lies mechanical support, and if the immateriality of these images and symbols gives rise to a new approach to the relationship between human being and object, the analysis will be one of the individual's connections with the material support underlying the new culture of immateriality.

The real problem, though, resides in the kind of relationship individuals will establish with the new material foundation that they will consider simply as part of the decor of the environment. Consequently they reject this from their own field of consciousness and thus forget it entirely for the benefit of the immaterial imago (which is slightly material in itself), which directs them: Its *credibility* is becoming more important than its *verity*. But since this same individual is constantly reminded of this material base by screens and terminals, the question is this: Under what conditions of reliability will the material foundation be truly forgotten and replaced by dubious, hazy, generalized images as only points of reference for a subsequent conscious activity?

These material supports reveal themselves only by their imperfections—by the necessity to plug in a computer, for example, and to turn it on in order to make it work. The infrastructure subsists in the fact that the elimination of the most insignificant contact—indeed, a telecontact—can extract all that the most advanced technology has put into a computer. The material supports remind us of our dependence on them, especially by such defects as the smudge on a poorly cleaned vinyl disk, the high-rise apartment building that interferes with radio reception, or the interfering lines that intrude in our private telephone conversations.

As an individual, here and now, I could forget all the hardware if only it were to reach perfection, something rarely attained in practice. Striving for that perfection and that absolute reliability is precisely the fundamental role of the one who *designs* the machinery of imagery, simulation, and communication; the standing of quality, without which no postindustrial world could subsist, is always questioned, as is the fleeting "perfect moment" of my relationship with the world.

The role of the designer, then, is not so much to create "new" objects to serve as structural supports of an immaterial culture as to insist on an environment of implacable stability. Before introducing something new, the designer must protect the status quo, which permits individuals to participate spontaneously and with little effort in the seductive immateriality of today's world.

REFERENCES

Adam, Villiers de l'Isle, *L'eve future, Tomorrow's Eve,* a science fiction poetical novel. Transl. Robert M. Adams, Urbana, Il,: University of Illinois Press, 1982 (Paris: Editions 10.18, 1962).

Alexander, Christopher, *Notes on the Synthesis of Form.* Cambridge, Mass: Harvard University Press, 1967.

Bell, Daniel, *The Coming of Post-Industrial Society.* New York: Basic Books, 1973.

DeForge, Yves, *Le graphisme technique.* Seyssel: Champ-Vallon, 1981.

Moles, Abraham A., *Theory of Information and Aesthetic Perception.* Champaign: University of Illinois Press (Paris, Denoë, 1958).

Moles, Abraham A., "Vom Design des Gegenstandes zum Design der Umgebung." *Design ist Unsichtbar.* Wien: Locker, 1981.

Moles, Abraham A., "The Comprehensive Guarantee: A New Consumer's Value." *Design Issues* II, no. 1 (1985), 53–65.

Moles, Abraham A., *Théorie structurale de la communication et société.* Paris: Masson, 1986.

Moles, Abraham A., *Vivre avec les choses: contre une culture immatérielle.* Paris: Art Press, 1987.

Simon, Herbert A., *The Architecture of Complexity.* Pittsburgh, Pa.: Carnegie Institute of Technology, 1962.

Larroche, Hélène and Yan Tuchny, *L'Objet industriel en question.* Paris: Editions du Regard, 1985.

A Newfound Bareness on the Threshold of the Third Millennium

Maurizio Morgantini

On the threshold of the third millennium, we are venturing headlong into the desert of design. Contrasting approaches to the new physical role of objects and their progressive dematerialization have revealed the lack of a clear-cut vision of design. Despite efforts by historians and the marketing sphere—more often than not for vested interest—to string matters together, design seems to be evading the logic of the new course. We are now up against this contrast between a design utterly void of any real meaning—with a glut of forms and functions kept alive by a production sphere plagued with excess—and a new approach that for the first time welcomes the unconditional involvement of technology.

For a long time, this technology has been relegated to a somewhat antithetic role, compared to that played by "poetic feeling," and has often been confounded with technique. Only recently has technology begun to be identified with design, even though this phenomenon is taking place on two levels. The first is the communicational, epidermic, and predominantly visual plane, almost an expressive approach that dwells upon the prevailing vogue of high technology. The second level, which is infinitely more complex, identifies design with the technological project—in other words, with the selection and the control of the in-

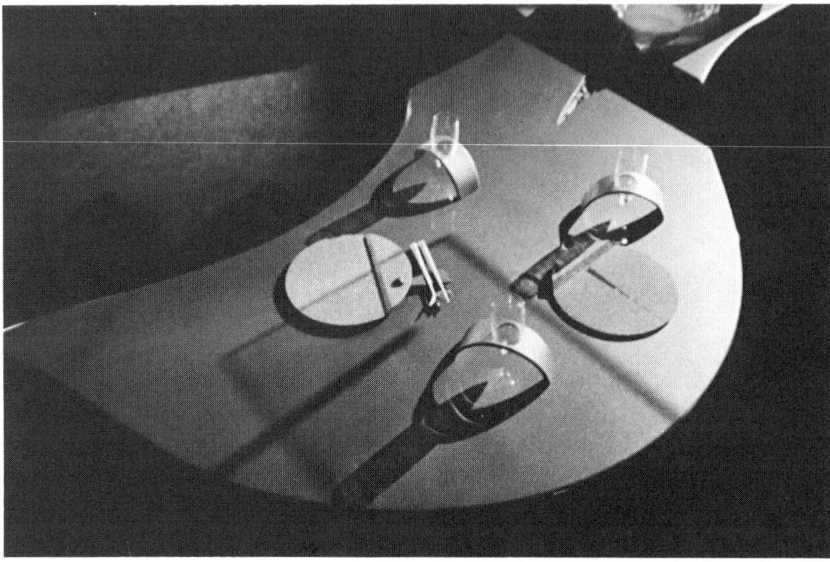

FIGURE 1. Workstation 1: detail of the reflecting zone, with objects for the ritual of eating.

numerable techniques behind this extraordinary synthetic framework. The second level is an existential condition based on continual option, on the complexity of methods geared toward perfection, and on the consequent crisis regarding purpose.

"Technology as an existential condition" is a proposition that is shedding the debris of a collective imagination kindled by robots, sidereal spaces, and food in tablet form. In fact, the history of design, the finalized evolution of the project, is predominantly a history of technology that has been put into effect, and its very realization is "communicative." Although a science does not necessarily have to communicate and a technique is such only if it is communicable, technology itself is pure communication. The technological project stands as the horizontal consolidation of vertical disciplines, crossing and blending divergent and departmentalized learning and playing a role of selection and organization across the fields of knowledge and awareness, transforming them into communication. A car, for example, is a combination of different scientific techniques and know-how: the internal combustion engine, the resilience of the materials used, the degree of reduced drag radio when driving, and the electronics of the controls. The combination of all these is a technological synthesis that brings the various degrees of know-how onto the same wavelength, enables the project to function, and conveys (therefore communicates) the possibility of application.

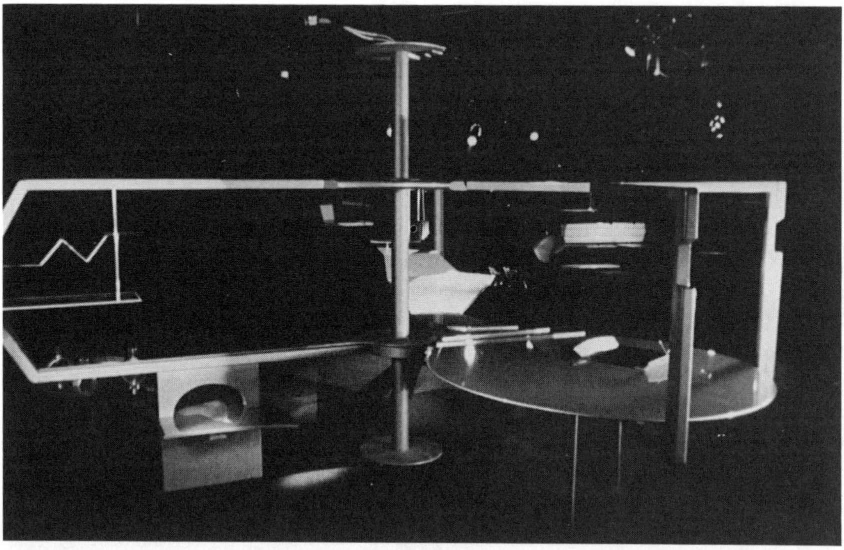

FIGURE 2. Workstation 1: this work island is built around a telematic tree; it has three zones: working, communicating (the round table), and reflecting.

Technology, as does politics, calls for the consideration of all possible hypotheses and the highest degree of teamwork when handling a project. Technology is, therefore, the liquid state, the absolute zero of political theory. For this reason, we are wrong to look upon technological development as merely the widening of the sphere of application and the increase of instrumental quality. This definition is void of all reason and purpose: It is a misconception and is nothing but the consequence of excessive rhetoric as it pertains to the expressive language of architecture and design, nothing but the perpetuation of a discrimination between art and technique. The "high-tech/high-touch" formula, often used as a means of balance, as a way of laying stress on opposing forces, and as an antidote to alienation brought on by the application of advanced technologies, falls down because it lacks substance. The formula seeks the equation of the zero setting—minus 100 plus 100 equals zero—the neutralization, in other words, of the computer (of its alien presence) by means of using pastel shades for the screen.

More interesting and, indeed, more worthy of attention, on the other hand, is our imagining that advanced technological society will be a ritual society. But ritual is neither the unforeseen counterpart to what is artificial nor its antibody. The ominous formula of the chemical cure makes way for this point of liaison between advanced technology and the memory of ancient mythology, the closing of a circle that relates the

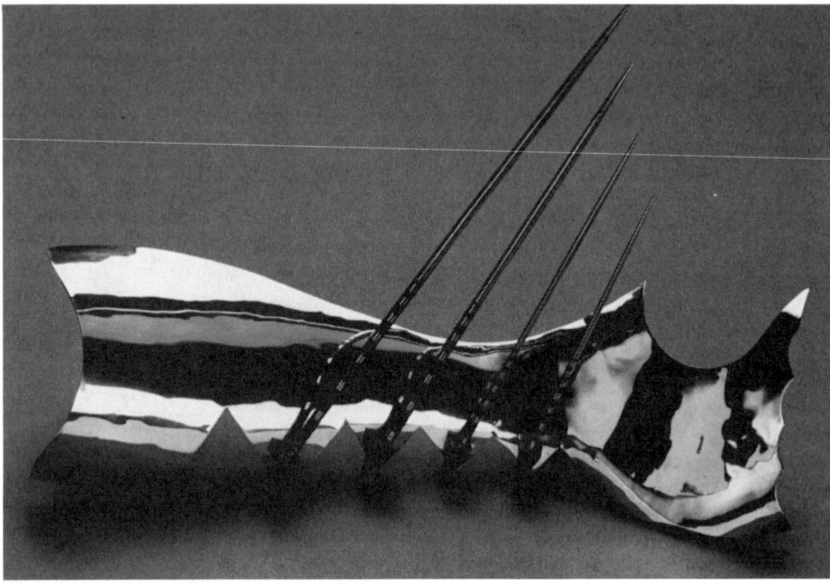

FIGURE 3. From the exhibition "The Emperor's Office," NEOCON, Chicago, June 1987: arm synthesizer, an electronic musical instrument following the shape of the arm as a double skin.

entire project, from its remote oniric origin to the current reappropriation of the qualities attributed to objects.

The ability of humans to fly today is not just the result of an invention a century ago but stems from a dream "as old as the hills." The dream developed into a myth and carried through history until relatively recently it became reality. This connection between superhuman powers, at one time attributed to the gods, and the newfound faculties that have been conferred upon us by technology could well serve as the basis for the new and different design potential of the third millennium.

The current faceless glut of objects, the multifunctional crossbreeding (watches with calculators, or calculators with watches), the loss of object identity and the fact that objects are no longer classifiable individually are counterpoised by the sense of a more mature and accepted artificiality. The liaison between natural and hyperartificial is already under way. The object of the third technological generation, electronic intelligence, is realized in a way that is more organic than mechanical. The interfaces are closer to us. The linguistic expression of computers is becoming human and building up a relationship of symbiosis, allowing the most natural acts to bring about changes and make something happen. For example, the voice is able to substitute for the keyboard or a

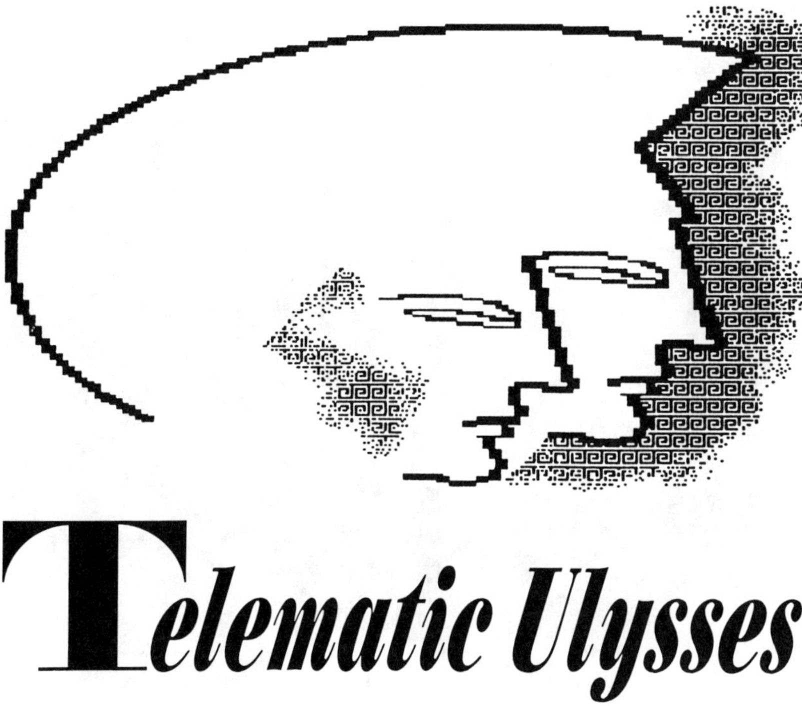

FIGURE 4. Telematic Ulysses: Logodesign by XENO, Chicago.

sensor is able to recognize the palm of a hand waving across it, instead of using magnetic cards. *We* are the gods we thought we had lost touch with: gods who set something in motion and made something happen because of some mysterious and invisible force. All that could be seen was the initial act and its effect, the ritual action and its consequence. These mythological gods and this projection of our dreams were perhaps the first indication of a process of design that, through material objects, would eventually become concerned with the body once again. Every action will be more akin to a ritual: A state of creative awareness and of collective narcissism will bestow upon objects and environments the disposition of a newfound sacral character. Will the objects of the third millennium be visible or invisible? Will they still be remote from the body, close at hand, or even part and parcel of it?

Hyperobjects are similar to jewels, and they stem from the blending of material and immaterial forces. They will indeed be able to take their form from those objects that have known how to wend their own way across the pages of history, gathering their charge of symbolic, evocative, and mythical values. Carpets, walking sticks, rings, pens, hats, arms, and

FIGURE 5. Video-disk book designed by Peter Langmar.

jewels will once again express applicational functions that lie somewhere between metaphor and technology, between what is magical and what is possible. The keyboard of a computer still stems directly from the angled, mechanical keyboard of a typewriter; although flat keyboards, which allow the fingers to glide across the surface, could be developed from a technical point of view, there is considerable resistance from those who have used typewriters, and, therefore, the flat keyboard has only partially been adopted. A century ago keyboards did not even exist, but we have grown used to them and this habit is what restrains further change. Strangely enough, it seems that the hand is reluctant to acknowledge a wide, more sophisticated span of feeling, limiting itself to the range a violinist is capable of with his instrument (and perhaps far less, when you think about it). It will, therefore, be up to the hyperobjects, the new interfaces, to inspire new uses for the body. Just as the keyboard could go out of circulation, to be substituted by digital sensors, the "mouse" on personal computers could well disappear in the same way; both can be substituted by rings that record every movement of the hand in a given space. The conventional and repetitive nature of mass-produced objects will make way for the full intensity and beauty of gestural "body language," a motivating gesture that can be taken back over the pages of history to its remote, mysterious, and poetic origin as an emotive gesture. A gestural act, therefore, is ritual in the same way that music is. Likewise, the instruments that generate the harmony are both remote and close at hand.

This new condition of bareness is but a metaphor, not yet developed, of an imminent evanescence of objects, the functional metamorphosis of which is already fully under way.

The Technologies of Self-Fashioning: Beyond Universality and Variance of the Industrial Product

Tufan Orel

Translated from the French by Susan Willis-Altamirano

To understand objects, one must ask the following questions. How does the user experience the world of industrial products? To be more precise, what kind of evolution of daily experience has taken place in the world of industrial products from the beginning of consumer society to the present?

MILIEU AND THE FUNCTIONAL CIRCLE

The notion of experience calls upon both subjective perception of objects and concerted actions for the accomplishment of real tasks at the time they are carried out. When I designate a piece of wood or steel covered with material as a seat, it is not sufficient for it merely to bear this meaning. To become an actual seat, it must also fulfill the requirements of seating. Furthermore, this seat must be a form of seating to which I have become accustomed. In other words, it must fulfill the function of seating. It is, therefore, not sufficient to be attentive only to "systems of meaning" that refer to objects; one must also take into account the satisfaction they provide for the diverse needs of real, habitual action,

Une télécommande sur baladeur. J'en rêvais.
Ne plus sortir de mon bocal. Pouvoir plonger au plus profond du son.
Enfin ! Je me sens des ailes.
AUTO REVERSE
AIWA
Voici le premier Baladeur à télécommande : le HS-J08 Aiwa. Le premier lecteur-enregistreur stéréo avec Tuner AM/FM. Auto-reverse, Dolby, mécanisme anti-roulis, position Métal. Touches à effleurement et senseur de sélection musical. Le fin du fin du son. En 265 g. Et il y a 16 autres modèles de Baladeurs, chez Aiwa.
AIWA®
Quand on devient Aiwa, on le reste.
Aiwa France. 117, rue d'Aguesseau. 92100 Boulogne. Tél. 46.04.81.90

FIGURE 1.

within the framework of a given culture. To speak of experience implies establishing a principle of adequacy between perception and action, given sense and accomplished act. In brief, the *sense* ought to impress a real function on the *milieu* of which industrial products are a part. Above all, the term *milieu* must not be confused with the term *environment.* The latter consists of considering the world of industrial products as an environment presupposing an exterior world in which one acts on subjects and obliges them to interiorize its technical, moral, or political criteria. Milieu, on the other hand, tends to imply a mutual adjustment of the world of objects to that of the user. Therein is a harmonization or natural composition of object and subject, such as that of bee and flower. It is in this light that one can consider, for example, that the constituent fittings of a house, as they have evolved since the nineteenth century, are a vital milieu or even, to borrow a term of a well-known ethologist, Uexküll, a "functional circle."[1]

The idea of the milieu, or of the functional circle, keeps us from the two most common forms of reductionism in the study of industrial products:

1. The thesis in which we consider industrial products as having an intrinsic logic of their own and see a sort of teleology in their evolution. This approach is more well known under the name of the *machine kingdom.*[2]
2. The thesis that aims to establish a moral code of styles of objects and assumes that it is possible to pinpoint a certain number of "systems of meaning" or modern "mythologies," which are likely to enlighten us on ideological values hidden in systems of objects.[3]

In the first case, a historicism of the functional product is favored, and little or no attention is paid to the attitude and reaction of individuals toward the technological world. In the second case, the role of the marginal value of the product is accentuated, and there is a failure to specify by which technological ontology each style is supported.

The idea of a functional circle does not necessarily conceal a new trap of absolute functionalism: An armchair that molds itself perfectly to the form of our bodies does not mold itself to the form of an absolute body but to a body resulting from its participation in a *certain* functional circle, situated historically. In this respect some theorists of human action, such as Abraham Moles, are certainly right to insist on the changing relationship between acts and industrial objects in modern society. But the theory used to explain these changes is not always satisfactory. In order to understand the new modifications of acts, we seem instead to need a modification of the theory of action, as will be argued later. As for the famous distinction between the *form* and the *function* (or materiality) of the product, we agree with Uexküll that when one changes "from one milieu to another, it is not just the meaning of an object which changes, but also the structure of all its features, both formal and material."[4]

After these preliminary remarks, we have to consider whether or not a new functional circle has come into existence, one which is in antagonistic opposition to the one we have known since the beginning of consumer society. Has the world of industrial products modified our experiences *in an essential way*? If so, what objects and actions have been responsible for it?

THE INDUSTRIAL PRODUCT

Disregarding the sociohistorical factors that have conditioned the birth of the industrial or serial product, one must nevertheless provide some details about the genesis and evolution of the industrial product so as to render its definition at the beginning of the consumer age more intelligible.

The idea of a model and series originally dates from the Renaissance. As early as the fifteenth century, engineers and artists referred to a *disegno* to evoke the quality or esthetic line of a picture or model. The relationship between a model and series is primarily one of a principle of repetition: serialized reproduction of the initial model. But it is also the relationship between "theme and variations"—the possibility of reproducing the same model and adding combinatory elements.[5] However, it took until the age of mass production to realize technologically what had previously been of a speculative or artistic nature. Under the "force of the division of labour," the industrial product becomes the equivalent of mass production. This is demonstrated, for instance, in England, by Josiah Wedgwood, the specialized pottery manufacturer, who in 1769 was already paying his modeler or designer twice as much as a simple craftsman and who sought to eliminate variations in production in order to achieve perfect standardization.[6]

After the mid nineteenth century, the industrial product entered the sphere of influence of mechanization. At a time when "mechanization takes power," as Giedion, the historian of everyday objects, puts it, the industrial product became the constituent furnishing of houses and offices.[7] That is how the beginning of the technological milieu came into being together with a new functional circle. This new milieu survived despite the revolts and resistance of such movements as Arts and Crafts (instigated by William Morris and Matthew Arnold, who wanted to restore the old symbolic and moral order of the artisan of the Middle Ages along with the natural harmony of ritual and tool) and in spite of William Pater's and Oscar Wilde's exaltation of esthetic values and the versatility of the Renaissance artists.[8]

After this classic age of the industrial product, a new age, often referred to as the "takeover of design," began in the early twentieth century. The change in direction was decisive: Instead of the opposition to, or at most, the timid art-industry relationship that existed at the end of the nineteenth century, one found a fusion, or rather the reconciliation of producers of the technical field ("technosphere") and artisans of the esthetic sphere. The aesthetic ideas that bubbled up in the industrial world between 1910 and 1950—art deco, de Stijl, Vorticism, Futurism, Bauhaus, and others—gradually combined to form the reality of the industrial product as consumer society knew it at the beginning.

The interpenetration of the object of design and the technical product has not of course succeeded in dispelling all the problems concerning the status of applied arts as far as industry is concerned. The problem of "style" has been left in abeyance and can still be thought of as opposed to the industrial reality of "design."[9] Whatever the case may be, the industrial product takes on what one could call a *matérialized formalism*, even if many avant-garde debates are still beyond its area of influence. In order to illustrate what was considered as the industrial product toward the beginning of consumer society, it is sufficient to refer to the enlightening remarks of Gropius, the first director of the Bauhaus, on the subject of design: "The term design generally embraces the entire orbit of what surrounds us and what has been created by man's hands, from the simple object or everyday utensil to the complete layout of an entire town."[10]

CONSUMER SOCIETY AND THE UTILITARIAN OBJECT

What essentially characterized the beginning of consumer society was the acquisition of industrial products of universal value and the subsequent generalization of these constituent fittings of our everyday milieu, into the home, the office, and the public place.[11] It was a decisive period for the "natural composition" between products that had been making their appearance since the end of the nineteenth century and subjects or users. This natural composition is manifest in the ideas of the era. In this respect, one can speak of a consensual conscience of producers, users, and advertising executives to the world of domestic fittings. To form an opinion of their mental attitude, it would be sufficient to read the placards of the age of Réclame.[12]

If, on these placards, each product seeks to outdo the others in

FIGURE 2.

order to sell itself, it is not done to mask or conceal its real value and function. When the industrial product went up on the public stage in the public square, it did not boast more than it could actually do or satisfy. It is true, however, that the rhetoric of the marvelous sometimes mingles with the rhetoric of the functional circle. How then can one explain that? One must see the natural cycle of each functional circle in that. This natural cycle could be represented by three essential mental states of the users of the industrial object.

When users are drawn into the field of attraction of the functional circle for the first time, their joy at its discovery often manifests itself in fascination and euphoria, hence such an expression as marvelous. Later, there is a period of calm when it is taken for granted and has been recast by the user into a comforting object. At this point, the discourse on objects has turned utilitarian. Finally, the functional circle begins to be boring: What was initially *tonic*, then *calming*, has now become *toxic*. And it is perhaps at this point that utilitarian objects take on a "marginal symbolism," disguising themselves as objects of desire or of cult. But it is by no means obvious that this cultural or "sacramental" respect toward industrial objects succeeds in displacing the persistent constituent milieu; the breach must be sought elsewhere by asking the key question: Have the material *function* of objects and their accompanying *actions* been radically modified? To come back to what is the marvelous element at the beginning of consumer society: If one must really look for a mythological motif, it would be more appropriate to speak of the pragmatics of the marvelous. Just as in the antique world the sun, the moon, fire, and fountains were venerated as gods because they were life-favoring elements (according to Prodicos of Ceos), the exaltation of utilitarian objects can doubtless be explained by the real satisfaction they provided in making life easy.

DIVERSIFICATION OF THE INDUSTRIAL PRODUCT: FROM BOURGEOIS TASTE TO STYLES OF LIFE

It would certainly be a great mistake of appreciation to think that it is sufficient to evoke the feelings of comfort and well-being that industrial products elicit for a person to apprehend daily experience in consumer society. We certainly cannot ignore the fact that as soon as the latter had begun to function, or be manifest, a consciousness of style evolved. This consciousness allowed the creation of a market of distinction that, by diversifying products, gave rise to "conspicuous consumption."[13] In connection with this, and at a more technical level, the diffusion of products that participate in the functional circle, that is, serialized products, is doubled by what Baudrillard calls a "second seriality."[14] In other words, objects that are useful to our lives simultaneously become objects of desire or cult.

It would be impossible for us in this case to draw up an exhaustive taxonomy of the entire number of objects of desire or cult. Let us nevertheless distinguish between two fundamental categories:

- Objects of desire considered as a variation on a model. In this case, the object remains organically linked to a matrix design, to what we call a *materialized formalism.*
- Objects not directly derived from a matrix design and that one can call gadgets or playthings.

Leaving aside the second category, we shall occupy ourselves exclusively with the first. But, as the origin of the diversification of the industrial product and the market of distinction is not to be found at the beginning of consumer society, it would be fitting to give some details of their evolution since the last century.

A short time after the industrial product was drawn into the field of attraction of mechanization, variation (which one manufacturer had tried to eliminate throughout the entire manufacturing era) became a new technological and economic objective. The industrial philosophy of the era of mechanization was to "combine standardization and variation." It is of little importance whether what lay behind this new technological aim was the diversification of needs (the bourgeois argument) or the whims or talent of stylists (the esthetic or stylistic argument). What is certain is that it allowed the development of an outer symbolism of objects parallel to their inner cultural value: The culture of the chair would be coupled with a symbolism of the chair, the bathroom culture with a bathroom symbolism, and so forth. This practice pervaded all sectors of industrial production. For example, from 1895, Montgomery Ward, a firm specializing in the manufacture of pocket knives, was in a position to offer its customers 131 different types of products derived from a single matrix model and based on diversification according to the criteria of age, sex, and status. Between 1890 and 1900, in the catalogs of the navy stores in the United States, servants' and masters' cutlery were differentiated. Similarly, between 1933 and 1939, E. Gomme, an English furniture firm, was to produce forty-seven variations of sideboards from ten matrix models to appeal to different customers' tastes.[15]

Not until the 1960s did the market of distinction become more widespread. Object symbolism first imposed by bourgeois taste, later greatly influenced by pop art and pop culture, was overtaken by a symbolism of the "futile" or of "consummation."[16] The symbolism of the "futile" was taken into account very seriously. It can be characterized by an attentiveness to the "message" of the industrial product and a disregard for its material content. This symbolism is, indeed, fairly characteristic of the field of advertising, as opposed to the era of *réclame*, the era when the *message* seeks to associate a product with a *meaning*. Semiologists, the grammarians of style of the post-1968 industrial product, often disregard the real function of objects and speak of a *mythology* of objects of desire and of objects of "meaning." Their theoretical contribution could be summed up thus: Beneath objects of comfort hides the imaginary.

The most recent phase of the market of distinction is characterized by what are called *styles of life*.[17] The criteria of distinction are basically founded on the sensitivity and values of either microgroups or the plurality of *Lebenswelten*. This phase is a very good illustration of the "supreme stage" of diversification of the industrial product and of the market of distinction since the mid 1970s.

THE MARKET OF DISTINCTION AND THE FUNCTIONAL CIRCLE

We can therefore see that there are futile as well as useful products in consumer society, an established functional circle but also esthetic-imaginary spheres of life-styles. Can we then conclude that there are two systems of products and two types of experience that coexist? Far from it. One must see the problem from another angle. Let us try to detect the contradictory links between the two and to uncover their essential relationship. We shall try to establish whether their relationship is one of symmetric opposition or of complementarity.

In order to shed more light on this matter, one could carry out research on two different fronts: (1) the internal contradictions of the market of distinction or (2) the external contradictions, that is, its relationship to the functional circle—particularly at its supreme stage, which vehicles the ideology of life-styles.

INTERNAL CONTRADICTIONS

In fact, at least at the beginning, the market of distinction never hid its true ideological intentions. Leaving aside the idea of comfort, which must be linked to the functional circle, the marketer always promoted luxury when he or she could not designate anything other than differences of class or status. It is a well-established fact that the bourgeois model was the first cultural model of consumer society, as this model was at the origin of the social dilemma on consumption. One either had to "imiter le bourgeois" or "épater le bourgeois" (imitate bourgeois values or shake them). But this market also had recourse to another, more ambivalent notion of personality or personalization. If we really are to look for the major contradiction of the market of distinction, it is no doubt necessary to look more deeply into what is behind this term, which has often served to form such paradoxical slogans as "personalize your living surroundings."

The critics of consumer society, such as Baudrillard and, more recently, Adrian Forty, have already looked into this paradox.[18] Let us simply recall the terms. On one hand, the consumer is expected to be spontaneous, creative, and master of his or her own tastes, or even (to use one of Penny Sparke's coinages) to make "taste mistakes."[19] On the other hand, models, taste, and style norms are imposed on consumers so that they can personalize their own lives' surroundings. All in all, the freedom of the consumer is reduced to choosing already existing "archetypes," psychological and esthetic variants of "types," or model matrix techniques in order to forge herself or himself a personality. It is a remarkable fact that in this case it is not a question of creating or inventing one's own personality but of adopting or choosing an archetype of personality available on the market, or simply of forging oneself an archetypal personality.

Is this paradox still with us? Does it persist, as Adrian Forty seems to suggest, or is it disappearing because of recent transformations in consumer society? In order to answer these questions in a satisfactory manner, it is necessary first to look into the complex relationship between the market of distinction and the functional circle, and then into the problem of the "variance" of the industrial product.

EXTERNAL CONTRADICTIONS

Styles of life, which represent the supreme stage of the market of distinction and differentiation of the industrial product, are not destined to create a new milieu. They simply allow the permutation of stylistic or esthetic games within the sphere of constituent fittings such as have been universalized since the beginning of consumer society. At best, these products or fittings serve as a cathartic aid to their users or as a performance in esthetics, but they are not modified themselves as products. Neither the technodesign texture—its *materialized formalism*—nor the comfort they provide are altered: An outdated bathroom always retains its cultural function, one of regeneration.

It is possible in this case to consider the experiences—or rather the sensitivity of the user toward these products—as an islet of thriving existence that is then submerged in a more enveloping ontological framework represented by the functional circle. We shall call this sensitivity *a marginal transcendence of the vital or functional product.* The phrase implies a particular perceptive attitude toward products, which consists of regarding them as sacred for a moment (the duration of that moment being restricted by the duration of a fashion) with the aim of achieving esthetic life forms or life-styles. The sensitivity is a sort of sacrificial practice of the function.

Basically, it is of little importance if the users themselves mix and match the life-styles or the archetypes available on the market or if the producers themselves compile these assortments (which can in fact be referred to as "sets of archetypes"). One sure fact is that objects of desire or cult allow the satisfaction of esthetic perception that is contained within the perception and vital action of the functional circle. And if the constituent objects of our technical milieu allow themselves to be taken for what they are not, it is first and foremost because they are assured of having a *cultual* function. They can therefore allow themselves the luxury of also having a cultural function (*fonction culturelle*).

Could there not be other ways of conceiving the status of objects of desire and cult? Most certainly. Nothing prevents us from seeing their use as attempts to throw off the inertia of the existing functional circle that has tended to become a dysfunctional circle. But it is to be feared that these attempts will be made in vain. For styles of life do not begin with something to sit on but with the comfort or cultural postures of the chair; similarly, life-styles do not begin with a place to eat but with the constituent fittings of the kitchen, and so on.

The functional circle therefore encompasses all possible life-styles, and these styles do not create a new way of sitting, a new way of bathing, or other actions. They therefore display a transitory *sensitivity* and do not make outdated the constituent *experience* with our vital milieu. (In this respect, we should recall a clear-sighted remark of Giedion's: The shower, the sauna, and the bathtub are all part of the culture of "regeneration," which is, consequently, diametrically opposed to the culture of "ablution" that came before the age of mechanization.)[20]

THE BIRTH OF A NEW FUNCTIONAL CIRCLE?

If life-styles are not the necessary lever for shifting the functional circle established at the beginning of the consumer age, are we then to think that we will always be caught up in its area of influence? A few precursory signs indicate that such is not the case. Let us try to establish whence came the cracking sounds of this functional circle.

VARIANCE

During the last few years, both the conception and the production of the diversified industrial product have changed radically. As we have seen, the beginning of the consumer age coincided with the takeover of design, whereas today the industrial product is beginning to be dependent on the industries of artificial intelligence. Thanks to numerically controlled tools, robotization, and computer-aided conception, the close relationship between mass production and diversification is profoundly changing in nature. In this new industrial arrangement, the engineer-designer will be able to draw up a catalog of combinations, or the combination of combinations (*variance*) of the products that have not yet been produced. Then, when the client demands an individual object, the engineer in variance will command the entire process of production through a given code to produce the desired object.[21]

The modern idea of the diversification of the industrial product, that is, of variance, allows for the manufacture of nonserialized products. As the Italian designer Gaetano Pesce has already proposed, by introducing random touches into a series it is possible to create more than a thousand seats, all different. Nearer home, the idea is applied in a General Motors project. The Saturn project has the aim of bringing out nonserialized cars on the market, on one hand using methods of random production and, on the other, taking individual clients' suggestions into account. Furthermore, the possibility of working on the basis of an individual customer's orders is not without advantages in service and stocking. The product could be sold directly without having to store it, thus omitting the intermediate steps.[22]

Parallel to this idea of variance, other conceptual innovations, which are not unrelated to this problem, are coming from designers themselves. It is a question of making the designer a participant in projects of personal creation. In fact, instead of creating models and variances from a design and planning department, the designer can play the role of an "orientator" or of a pedagogue "for the self-production candidates" (those who want to mix and match their own living space or surroundings instead of submitting themselves to life-styles chosen by the market). The idea was introduced in 1974 by Enzo Mari ("A Proposal for a Personal Project") and, according to Raymond Guidot, consultant engineer at the Centre Beaubourg, it has a great future for the creation of personalized interiors.[23]

Are we to conceive of these examples—which will no doubt be far reaching in the coming years—as resulting from the technological demands of the robot age and from the talent or whims of designers? Perhaps. But we can by no means ignore the aspirations and current beliefs of the users, who certainly play a determining role for the favorable reception of these innovations.

The ideology of life-styles principally addressed itself to the sensitivity of adaptable microgroups or fashion groups, whereas variance seems to be intended for individuals who wish to create their own living surroundings. The ideology of life-styles, which is proud of its typologies, continues to maintain the confusion that results from identifying the mental state of individuals with the collective mental state of its group of reference. On the other hand, the idea of variance allows individuals to have personal myths, that is, to be concerned with creating for themselves a personal space or with selecting their own life's surroundings, without being dictated to by the archetypes of the market of distinction.[24]

And yet one could legitimately have a few doubts about the technological evolution of variance. Although its sincere aim seems to be to help individuals to create their own space, the possibility of producing nonserialized products can overcome certain constraints of a technological order is not completely sure. For does not an object subjected to a multiplicity of variances lose its essential technical functions? As for pedagogue designers, as long as they do not become engineers of taste, their role could be of great use. But in a society in which creation is becoming everyone's business, one could ask why one must continue to listen to the advice of professionals.

THE TECHNOLOGIES OF SELF-FASHIONING

Certain theoreticians of the industrial product, such as Abraham Moles, have often defended the argument that every contradiction between systems of objects and society must be resolved by analyzing the acts and actions of the user. In order to render intelligible the recent changes in attitude and belief regarding the industrial product, we should adopt this very fertile and highly operative idea. And yet we cannot today be satisfied with yesterday's tools of analysis, such as praxeology, which essentially deals with rational and fundamental action (less effort, efficiency of movement, and so forth). Actions carried out in using a certain category of recent products are difficult to apprehend as being part of the concepts and axioms of the classical theory of action.[25]

To bring to light the nature or, to be more exact, the tendency of the action of the individual agent of our time, we have recourse to the hypothesis of the "activities of self-fashioning" in our other works[26] by insisting on the following fact. Parallel to the development of *instrumental* action and of *communicative* or *symbolic* action (intersubjective symbolic games),[27] there are also actions of *self-fashioning* directed toward ourselves

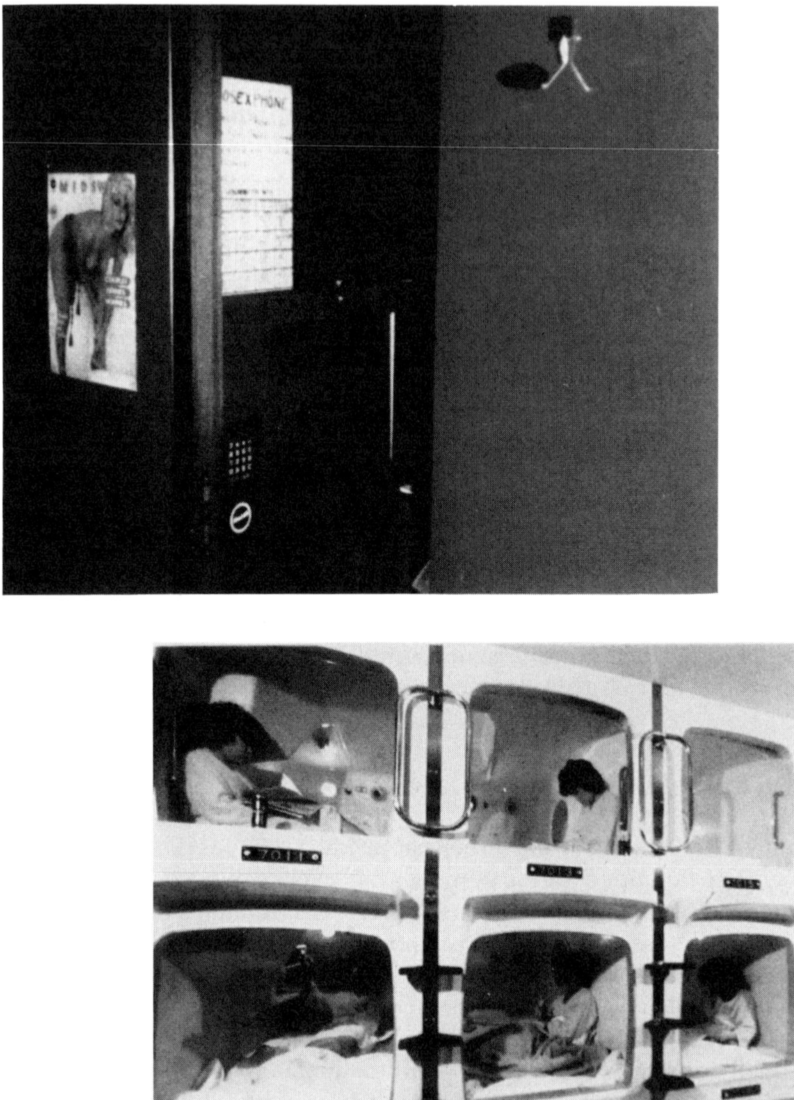

FIGURE 3a.

developing in modern society. These actions do not seek to forge an archetypal personality; instead they fashion or model our being, by carrying out experiments and research *on the self*. Though this type of activity could lead in the future to the creation of new social links and new social values, it is very difficult and far too premature to imagine

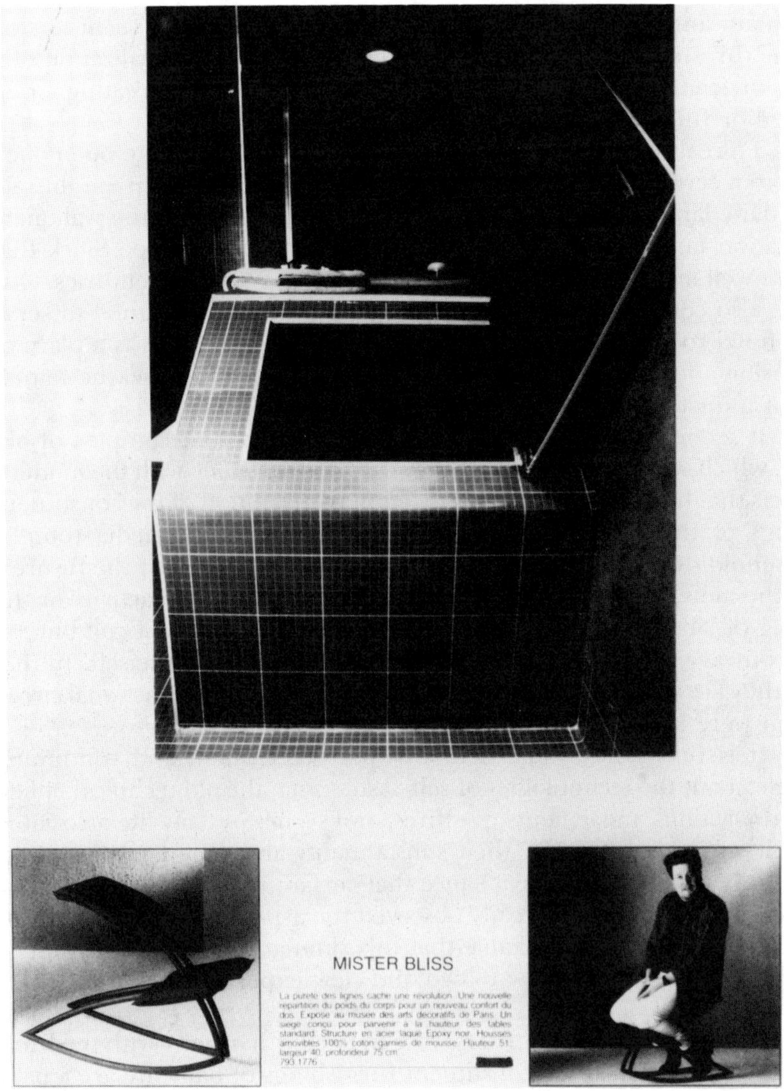

FIGURE 3b.

their nature at this stage. Nevertheless, the hypothesis of the "activities of self-fashioning" can help to shed light upon the new relationship that is beginning to establish itself between the world of objects and the action and experience of the users.

In a sense, the industries of variance could be considered an economic answer to this new aspiration of an "action upon oneself." But the main aim of the activities of self-fashioning does not seem to stop at the threshold of the market of variance. Beyond the realization of a *space* or framework of living, one finds the possibility of creating a new milieu or functional circle.

To realize that such a possibility exists, it is sufficient to open one's eyes to a certain category of complex objects that has begun to infiltrate our daily landscape. Among these objects one can count the walkman, floatation tanks, the "Mr. Bliss" chair designed by Philippe Stark that might well modify all the habitual postures of the last few centuries. One must also count the Japanese bedroom box, initially designed to serve as a hotel room but planned to be used soon in the house as a place of seclusion and meditation. We might mention also, of course, the soundproof individual TV cubicles of sex shops.

It seems obvious that one should not confuse this complex of objects, which we shall call the *technologies of self-fashioning*, with the modern objects that have developed directly from the tradition of the constituent fittings of the house: built-in media equipment (the "media room"), household devices, or even the equipment of the "house of the future." For the aims of the objects of self-fashioning are not interactivity or the saving of labor. They are destined to serve not as objects of cult but, on the contrary, as agents for carrying out experiments on the self. In that way they are symmetrically opposed to the objects of the functional circle, which were habitual up until now.

It is fully conceivable that with the acceleration and continuing beliefs about the technologies of self-fashioning, the old technical milieu and the actions, movements, postures, and scenes of daily life associated with it will completely lose their substantiality and degenerate as a vital milieu. It is perhaps not by chance that certain philosophers and social critics of our time have already observed the aspect of simulation of our social life. It is highly probable that this simulation is intimately linked to the emergence of a new milieu and new experience with industrial products.

At a time when Westerners were in perfect harmony with the functional circle of which the constituent furnishings of daily life are a part, reality was defined and maintained by supporting itself on the world of universal, uniform objects. In order to experience this world intensely, every individual fell back on objects of desire or cult through a game of identification to groups of reference and their esthetic and transcendent values. In order to satisfy individual aspirations, a reality of a second order had to be created by participating in styles of life. Today, however, the more the user/experimenter acts upon herself or himself, the more she or he tries things out on himself or herself (but also makes errors), the more she or he refers to a personal milieu by considering it as the only existing reality. Thus, the old technical and instrumental milieu and the entire complex of games and social rituals organically linked to it seem on the other hand to become a reality of a second order—a simulation.

Today we can see more than ever before that the old functional circle of consumer society was far from representing a final ontology. For, after all, though a person makes objects, he or she is also a demiurge and fashions his or her own reality. The individual is constantly in search of the most suitable functional circle.

Notes

[1]The term *milieu* or the *functional circle* corresponds to Uexküll's concept *Umwelt*. Abraham Moles suggested using the term *environment* as its equivalent, although he admits that this is not a fully adequate term. See Abraham Moles and Claude Zeltman, eds., *La communication* (Paris: La Bibliothèque du CEPL, 1971), p. 239 and Moles and Elisabeth Rohmer, *Théorie des actes: vers une écologie des actions* (Paris-Tournai: Casterman, 1971), pp. 33–69. We suggest using the term *milieu* because *environment* has a positivist connotation and is very much related to behaviorist epistemology, whereas *Umwelt* was a phenomenological term.

[2]The notion of *machine kingdom (Règne machinal)* was popularized in France by Guy Denielou, the former president of the University of Compiègne. See "Guy Denielou et le 'règne machinal' " (interview by P. Drouin), *Le Monde*, November 2, 1980. But originally the idea of studying objects in themselves—the ontology of technical objects—was suggested by the well-known work of Gilbert Simondon, *Du mode d'existence des objets technique* (Paris: Aubier-Montaigne, 1969). See also Yves Deforge, *Technologie et génétigue de l'objet industriel*, (Paris: Maloine, 1985) pp. 91–94.

[3]Here we refer to well-known studies by Roland Barthes, *Mythologies* (Paris: Editions du Seuil, 1957), trans. Annette Lavers (New York: Hill & Wang, 1972), and Jean Baudrillard, *Le système des objets* (Paris: Denoël/Gonthier, "Médiations," 1972).

[4]Jacob Von Uexküll, *Streifzüge durch die Umwelten von Tieren und Menschen—Bedeutungslehre* (Hamburg: Rowohlt Verlag, 1956). The French translation is by Philippe Muller, *Mondes animaux et monde humain* (Paris: Denoël/Gonthier, 1984) p. 160.

[5]On the subject of "Theme and Variations" during the Renaissance and its impact on design, technology, and literature, see Claude-Gilbert Dubois, "Artifice imité et diversifié," *Corps écrit*, 15 (Paris: 1985), pp. 105–16; "L'imitation sans limitation," *Revue de littérature comparée*, 3 (Paris, 1982), pp. 267–80; and *L'imaginaire de la Renaissance* (Paris: PUF, 1982).

[6]For a more detailed study of the Wedgwood firm and its designers, see Adrian Forty, *Objects of Desire: Design and Society 1750–1980* (New York: Pantheon, 1986), chapter 2, "The First Industrial Designers," pp. 29–41.

[7]Siegfried Giedion, *Mechanization Takes Command* (New York: Oxford University Press, 1948). The French translation is by Paule Guivarch, *La mécanisation au pouvoir*, 3 volumes (Paris: Denoël/Gonthier, "Médiations," 1983).

[8]On the industrial ideals of William Morris and Matthew Arnold, see Raymond Williams's classic study, *Culture and Society 1780–1950*, New York, Columbia University Press: 1983 chapters 6 and 7, pp. 120–61. As for Oscar Wilde, who was influenced in his early period by the Renaissance studies of Pater, his famous essay "The Soul of Man under Socialism" can be found in *Complete Works of Oscar Wilde* (London: Collins, 1969), pp. 1079–104.

We may also remember that the historian of technology Lewis Mumford later defended the ideal of "versatility" in his critiques of the modern technological world. *The Myth of the Machine: The Pentagon of Power* (New York: Harcourt Brace Jovanovich, 1967).

[9]Philippe Garner gives the following definition on the distinction between style and

design: "Style . . . may be seen as expressing the outward appearance of things, whereas *design* may be defined as a concern with the rational form, of which the 'style' is the changing face." See Stephen Bayley, Philippe Garner, and Deyan Sudjic, *Twentieth Century Style and Design* (London: Thames and Hudson, 1986), p. 7.

[10]As quoted by Giulio Carlo Argan, *Walter Gropius e il Bauhaus* (Turin: Einaudi, 1951). The French translation is by Elsa Bonan, *Gropius et le Bauhaus* (Paris: Denoël/ Gonthier, "Médiations," 1979).

[11]This is certainly a culturalistic and not an economic explanation of the origins of consumer society. To a certain extent, this distinction corresponds to the one Giedion has established between the constituent aspect of industrial products and their transitory aspect. See Giedion, *Mechanization Takes Command* Vol. II, chapter 4.

[12]*Réclame* as opposed to *message* is a technical term used by advertisers in France. *Réclame* comes from the first period of assistance to marketing, from the beginning of the century until the 1960s; it is a primary language of denotation, hyperbole, and magnification. The medium of the *réclame* is placards, whereas the *message* has been using multimedia since the appearance of TV spots in France in 1969. The *message* is based on the rhetorical figures of connotation. See Anne-Sylvie Patier, *Les discours publicitaires dans les films de lessive: essai de reconstruction symbolique* (Paris: Sorbonne-Nouvelle, 1986 [ronéo]).

[13]The notion of conspicuous consumption was first used by the sociologist Thorstein Veblen in a 1925 work to denote spending for the sake of prestige. It later became a common term to explain consumers' behavior in affluent society. See Veblen, *The Theory of the Leisure Class* (London: Allen & Unwin Books, 1970), pp. 60–80.

[14]Baudrillard, *Le système des objets*, p. 170.

[15]For these examples and many others, see Forty, *Objects of Desire: Design and Society 1750–1980*, chapter 4 and Penny Sparke, *Furniture* (New York: Bell & Hyman, 1986), p. 46.

[16]The term *consummation* was introduced into economic analysis by the French writer and philosopher Georges Bataille, *La part maudite* (Paris: Editions de Minuit, 1967), trans. Robert Hurley, *The Accursed Share*, 3 vol. (New York: Zone Books, 1988). Bataille opposed *consumption* to *consummation*. The former is based on the economic principle (usefulness, saving labor), whereas the latter is based on the spending (*dépense*) principle in the orgiastic sense. This distinction became quite popular among social scientists after 1968 in France.

[17]The origin of the expression *styles of life* certainly has a very profound philosophical and anthropological connotation. In this sense it can be compared to the term *Lebenswelt* (Husserl) or *Lebensformen* (Wittgenstein). Unfortunately, its trivial use in the marketing and design terminology of today has turned it into a term of caricature. It has become a catchall for multiple group fashions in a period. Obviously, we acknowledge in our analysis its use as a term of caricature. For a more recent review of "the styles of life," see François de Singly, "Etude de styles de vie," in *Encyclopaedia Universalis*, supplement 1987, pp. 401–05.

[18]See Baudrillard, *Le système des objets*, pp. 181–84 and Forty, *Objects of Desire: Design and Society 1750–1980*, pp. 106–07.

[19]Sparke, *Furniture*, p. 73.

[20]For further remarks on the differences between *ablution* (purification) and *regeneration* (relaxation), see Giedion, *Mechanization Takes Command*, Vol. III, part 7.

[21]Regarding the impact of new technologies on design, see Eric Gelman et al., "Wheels of the Future," *Newsweek*, June 17, 1985; Philippe Messine, *Les saturniens* (Paris: Editions de la découverte, 1987); Yves Deforge, "Objects among Thousands," in *The Handbook of Design Management* (Oxford: Basil Blackwell, 1987); and Danielle Quarante, *Eléments de design industriel* (Paris: Maloine, 1984). (*Note*: The term *variance* originally belonged to the language of probability: "The square of the standard deviation; the mean square of the deviations from the arithmetic mean of a frequency distribution" *[Webster's Third New International Dictionary]*. But today it belongs to the jargon of the consultant of combinations and denotes the possibility of creating unique objects through combinatory processes.)

[22]On the General Motors Saturn Project, see Messine, *Les saturniens*, chapter 6, pp.99–120. In spite of the evident originality of this project and its promise for future technology, the author states that since Fall 1986 the project has been jeopardized by the psychological resistance of workers to such an innovation, and by other problems. See pages 116 and 120.

[23]See Raymond Guidot's article "Des goûts et des 'ismes,' " in *Art Press*, "A l'heure du design," no. 7 (Paris: 1987), 58–61.

[24]The difference we establish between personal myths and archetypes aims at explaining the basic mental and ritualistic attitudes of the consumers to industrial products. This distinction could in a sense be compared to what William James termed the "religious experience," as opposed to "religion as an institution," *The Varieties of Religious Experience* (New York: The Modern Library, 1939). But in this specific context, we use these terms to refer to our field research on floatation tank experiences. Through this research we have discovered multiple personal myths of the users of this industrial product, which have nothing to do with the common opinion (trance, relaxation, body fitness, and so forth) related to it. See Tufan Orel, *Les technologies d'autofaçonnage: l'example des caissons à isolation sensorielle* (Brussels: Colloque de l'AISLF, 1985 [ronéo]).

[25]The close relationship between praxeology and design certainly needs special treatment beyond the scope of this chapter. But we can mention that all its defenders have adhered to a theory of efficient action: to control and master the exterior world. We observe today, on the other hand, that the users of industrial products try to act upon themselves through these products. Thus the problem is not to ask oneself if design can make good use of action theory but to question *which* action theory could best resolve its actual issues. On classical praxeology, see Alfred Espinas, "Les origines de la technologie," *Revue philosophique*, 15 (1890), 30 (Espinas was the founder of praxeology); Ludwig von Mises, *Human Action* (New Haven: Yale University Press, 1949 and 1963); Mario Bunge, *Epistémologie*, French translation by Hélène Donadieu (Paris: Maloine, 1983).

[26]See Tufan Orel, "A propos des techniques d'autofaçonnage," *Sociétés*, 1 (1984), 32–33 and "Travail et technologies," *Bulletin de l'AISLF*, 3 (1986), 77–80.

[27]The opposition established between the *instrumental* action and *symbolic* (or *communicative*) action has a very long intellectual history in the social sciences. The terms were introduced in the modern sense by the French sociologist Louis Weber, *Le rythme du progrès* (Paris: Felix Alcan, 1913). Later, the anthropologist Marcel Mauss popularized these two categories of action under the labels *ritual action* and *technical action*. More recently, the opposition *instrumental/symbolic* belongs to the vocabulary of post-modernism. In his critique of Marx's theory of human action, Habermas accentuated the predominance of symbolic or communicative action in modern society. See Jürgen Habermas, "Arbeit und Interaktion," in H. Braun and M. Riedel, eds., *Natur und Geschichte* (Stuttgart: Karl Löwith zum 70 Geburtstag, 1967), pp. 132–55; Daniel Bell, *The Coming of Post-Industrial Society* (New York: Basic Books, 1973); and Jean-François Lyotard, *La condition post-moderne* (Paris: Editions de Minuit, 1979). trans: University of Minnesota Press, 1984). We may also remember that the term post-modernism was introduced in debates on architecture in 1971 by Ihab Hassan. For a more detailed analysis of post-modernism in the context of this book see Chapter 1: "Postmodernism and the Deconstruction of Modernism," pp. 13–25.

Scribe 1984, or A Fiction in Design

Omar C. Iadin

From the capital of the centralized country, 22 March 1987

Dear Editor:

You have asked me to contribute a chapter for your book devoted to the design of the immaterial society, or the study of the new challenges that the technological revolutions are creating for the role, the definition, and the boundaries of design.

Surely I was, and still am, very flattered by your request. Indeed, I have been thinking of writing such an essay, you know, one essay with all the right footnotes, references, an introduction, and conclusion(s), preferably in a problematic or open-ended form. In order to do that I started looking into my files, my drawers, and my memory, trying to find the best material for answering your question. While doing this, I found a very large file in one of my offices bearing the title "Scribe 1984."

I have to confess that in that very moment, in a very short lapse, I went through various stages of amnesia. Having recovered the memory

FIGURE 1.

of that episode, I think that you might be interested in knowing why I could not and cannot bring myself to write the chapter you requested for your book. In many ways the answer lies in that semiforgotten folder inscribed in a very mysterious way.[2]

Scribe 1984 is the name of a research project that a few years later appears to be loaded with meanings that I had not recognized at the time of my involvement. Right now I can see that it was a turning point in my career and in my understanding of the specific challenges of the profession of designer. As you know, and as I guess, this is one of the reasons that you asked me to write a chapter for your book. I am a social scientist[3] specializing in the study of new work technologies—let's call it automation, to be short—both in industrial and nonindustrial settings. In addition to that, and in order to be more prepared to face the complexity of the new problems, I also studied psychology, ergonomics, and human factors at several important European universities and research centers.[4] Until 1981 or 1982, most of my work had been purely academic (articles, books, diplomas, some teaching, and travel to conferences) with few and very distant contacts with the world of industry, or bureaucracies, or organizations, or whatever. The real world was not yet completely disclosed to me. I distinctly remember the end of my first day as a "consultant," although I do not remember exactly how it happened, or when. I was certainly tired and the day had been long, but I also had the sensation of being useful, of finally being recognized by a world that was "other than I."

From then on, my second career was fast and successful. I liked, or at least was puzzled by, the easiness of the switch and by the rapid sequence in which research, results, analysis, intervention, action, payments, and travels happened around me. I remember also the excitement of sleepless nights, early morning flights, and lengthy brain-storming sessions; the long list of names; and the variety of business cards I would receive after my various performances. Surprisingly enough, my academic career did not suffer from my "double life." To the contrary, I was almost encouraged to do more and, given that my list of publications was growing, I could not complain. After all, as your academic world says, "publish or perish." At the height of this wave, around the end of 1983, while I was working in the "centralized country" (as we might call it), I was asked to join a very select group of persons involved with the project Scribe 1984. I remember my surprise when I heard the name of this project. It reminded me of a certain Orwellian atmosphere; at least, that was my personal free association, largely because I wanted to do "something" to celebrate George Orwell's famous novel.[4]

Suddenly, I knew that instead of writing just a piece of academic research, I had the opportunity to work in a real setting, solving real problems of an unprecedented scale. But some additional information

FIGURE 2.

is necessary to explain the reasons for my excitement. In the centralized country, the balance of power had just shifted in favor of the party of Modernity. For the first time in almost fifty years, Modernity was in charge of a very important, rich, and large country. The rest of the world was observing this unique experience with mixed feelings and fears. But you know much more than I do about international politics and do not need more information on that episode. The project of the Modernity Party for the centralized country was, among other things, to be ready for the challenge of the future, to enter the third millennium, or the twenty-first century, well prepared to compete with the other countries belonging to the club of seven, and also to resist the rise of young countries. For a few years, it was to seem that no stone should be left unturned. Modernity and innovation through design were to happen everywhere, especially in the realm of the country's most famous specialty, that is to say the Bu Reau.[5]

The Keeper of the centralized country, also the leader of the Modernity Party, had decided that the most important and historically significant part of the Bu Reau, the Cas Hoffice, was not able to perform its operation in conformity with certain requirements for the new century. He called on his Gobet Ween and officially gave him the mission,[6] or the order (here the versions of the story differ), to redesign the Cas Hoffice in order to transform it into the most perfect example, the prototype, of the administrative and immaterial life of the future: clean, no dust, no paper, interactive, and open to all. It was to be an administrative Glass House completely grounded to the power of new technologies. The idea behind this project, I suppose, was to set the example, to renovate the most ancient one, and then to continue this politics of the symbol (as a friend of mine aptly called it). At the time, I must confess I did not understand very well what my friend meant. I do know now, but it is too late.

The Gobet Ween immediately instructed[7] his subordinate, Mr. Char Ge, who was responsible for the activities of the Cas Hoffice, that his ministry was to receive the privilege of being the first and the most important operation of the politics of Modernity by Design, as the operation was officially named. The Gobet Ween explained to Mr. Char Ge and his fellow citizens that in a few years, thanks to new technologies and the Modernity by Design, the face and many of the functions of the centralized country would be simplified, accessible, and easy.

To make a very long story short, after a while I was asked if I could be interested in becoming one of the experts in the operation of Modernity. Can you imagine my position and my excitement? It would have been an immense design operation, involving the competencies of dozens of experts in fields ranging from information technology, to interior design, to social science, to architecture. (A new Cas Hoffice was to be built, and the old redoubt was to be used as a museum: That gives you an idea of how old it was.) This, at least, was my perception of the operation. How else could someone, even someone so highly placed as the Keeper, decide to resettle thousands of employees of the Cas Hoffice, to change their environment and working conditions, to give them machines instead of quills, and to eliminate the walls that for centuries had sheltered them from the eyes of the "public"? This was the very best way to celebrate 1984 and to prove George Orwell wrong. Had the Keeper, who is a very cultivated man, also thought of that? I was not surprised. In a country in which the politics of the symbol was so central, all these implications must have been explored. And what an idea—to transform the old buildings of the execrated Cas Hoffice into The Great Museum, a lasting monument to the arts and culture in the emplacement of the Cas Hoffice.[8]

FIGURE 3.

The year 1984 started and I found myself totally involved with the project Scribe. Asked to form and lead a design and research group, I tried to gather together experts in as many specialties as I could, paying little heed to disciplinary boundaries.[9] I worked days and nights, I traveled, and I met researchers and designers. We prepared our action, our Modernizing Action. While I write this letter to you and I go through the heavy file, I remember many peculiarities of the Cas Hoffice that struck me almost immediately. At the time there were many more younger women, less powerful than their male colleagues, whose work consisted . . . Here, I suppose, I have to make a digression.[10] During the first period of my involvement in Scribe 1984, in my naive enthusiasm I had assumed too many things without asking. For instance, I had assumed that:

- The Keeper, the Gobet Ween, and Mr. Char Ge were all in agreement with the Modernizing Action.
- They knew the reasons behind the politics of the symbol that lay behind the Modernizing Action.
- The Modernizing Action was "desired" by most of the people working at the Cas Hoffice, from the bottom up, and there was a consensus.
- Everyone involved knew what was to be changed, why, and for what.

Only now, reading the notes and documents of my file, have I discovered elements that were already visible then, but not to me. As it must be clear by now, the main purposes of the Modernizing Action were to introduce and implement automation and to diffuse new technologies and new forms of organization throughout the most powerful, and one of the oldest bureaucratic structures of the centralized country. For that new "mission," a new site and a new building were chosen and designed at the request and under the major impulsion of the Keeper, and then of the Gobet Ween, and then . . . well, right now, I am not positive about the nature of the agreement of Mr. Char Ge and all the people working for Cas. In fact, during one of the numerous meetings that we had in order to prepare the Great Modernizing Design,[11] I remember asking the question in a very plain way, "What is to be designed, and why, and how?" Most of the faces around me looked surprised and bothered by the lack of tact in my question. Misinterpreting these signs, I became more specific and told them about the awkwardness of Change by Design, the enormous limits of technology, and the necessity of understanding the needs of the thousands of persons working in the various Cas Hoffices. In fact, I concluded with the question, "If you want to modernize, which part of your work has to be maintained, improved, or abandoned?" The meeting ended almost immediately after the question and everyone left at a remarkable speed. After every meeting the list of questions on one side grew longer and the silences on the other side increased. An uneasy situation began to overshadow the initial enthusiasm.

We were told that the building that was to house the entire Cas Hoffice was already under construction and the plan and the project had already been chosen directly by the Keeper, mostly for esthetic reasons, I suppose, given that no one, not even the team we had established, knew what was going to be put inside. Technologically speaking, the centralized country did not have a truly competitive edge in the domain of automation that was to be implemented with the Great Modernizing Design, but national pride and a certain xenophobia, prevented any further discussion of the subject—or any useful one, at least.[12] We were becoming even less active than at the beginning. The various multidisciplinary specialists had long conversations among themselves. Some of them liked the new offices, I mean the offices that were in the old office building that was to become the Great Museum . . . well, I am sure you understand.

A very sharp contrast was visible between the level of the discussions that we were having with the various persons at Cas and the public debate that was simultaneously taking place in the press, in parliament, and in the unions. Deontologically we were to be more or less silent and very reserved on the subject because we had access to sources no one else had, supposedly.[13] In fact, very often we were the last to be informed.

Because this letter is already too long and much of my file will never again be used or mentioned,[14] let me go directly to the end of my story. The Great Modernizing Design was becoming a very impotent exercise in patience and bureaucratic diplomacy. Our team was doing no specific study and was certainly not involved in any of the Great Modernizing Design decisions—not for the building, not for the technologies, not for the organization, even less for the employees. Finally one day, during one of the meetings, the former director of the OMC informed us that some specialists of the Vase Line Bureau of Methods were going to assist the Multidisciplinary Design team in minor points of detail. We had no reason to worry, he added, because we were the experts they needed. The Vase Line guys were there just for a few days. . . .

What else? I went back to academia, trying to work to understand what a science of design could be in a complex, fragmented, and technologically advanced society. Of the team, some are now rich and famous, some are just famous, and some are in between. We are old friends, even if not good friends. The Great Museum is still under construction. They are trying to put a circle in the square in front of it. The new Cas building is also under construction and should open a few years from now. . . .

Yours,
Omar C. Iadin

Notes

[1]I wish to thank my friend and colleague Stephen L. Rustow, AIA, AICP, Associate, I. M. Pei & Partners, for his kindness in providing the photographic maquettes of the Grand Louvre, a project very similar in scope and design to the one described in these pages.

[2]For deontological reasons, I have changed the names of the persons, organizations, places, and countries involved in my story. I do not have a specific reason, as you will discover in reading my letter, but I feel freer in doing this because, in our fields, where consulting is a large part of our activities, naming names is not yet a diffused practice. But does it really matter? Would we really know more knowing the identities of all the actors involved in my story?

[3]"Social scientist" is a term that conforms to American standards; elsewhere my specialty is called human(e) science. A funny linguistic puzzle, isn't it?

[4]Well, we can hide only a part of my story; but to say that I am European and that my story takes place in Europe does not reveal much, even though it seems to me your choice of contributors for the book has placed a specific role on this polarity, or opposition, between the two sides of the ocean, even three, if we count Japan. But, after all, you are European, too. . . .

[5]This idea required no imagination at all. In 1984 almost every university and journal in the world was organizing "something" to celebrate *1984*.

[6]Just think of Marx's *The 18th Brumaire of Louis Bonaparte.*

[7]The decree was published in the official journal of the centralized country.

[8]With another decree published a little later in the same official journal.

[9]I am sure that you have heard all about The Great Museum.

[10]The new technologies of information and the "science of design" seem to have no clear definition. Can there be one and, in this case, only one? I am quite sure that most of your contributors will, in one way or another, address this dilemma. For my part, I confess confusion.

[11]But this entire letter is a digression. I am taking all this space to tell you that I cannot answer your request and I am not really telling you why. Digressions are a dangerous tool for exchanging ideas. People get lost sometimes.

[12]Most of the meetings were informal because a decree was also in order to launch the Great Modernizing Design.

[13]The funny thing, as I recall it now, is that the Director of Cas in Charge of Technological Innovations had just left the well-known OMC, that is, the Office Machine Company, the famous American multinational, after twenty years as head of marketing operations.

[14]This could be the subject of very interesting research. What is the reason of the experts? Why is secrecy so often used even when it comes to perfectly pacific subjects? Why do experts accept such conditions?

[15]What could be the use—a study in missed alternatives? Roads not taken? It is already done, anyway. . . .

Expanding the Boundaries of Design: The Product Environment and the New User

Victor Margolin

While debates about the comparative merits of modern and postmodern style, Braun and Memphis, High Tech and High Touch rage worldwide, more radical changes in designing are taking place. Most significantly, microprocessors are making possible more complex products with a widened range of functions, resulting in a new flexible relation to the user, who selects specific functions from a broad set of options by programming the product or making choices from a visual menu.

Consider the capabilities of the videocassette recorder. A person can have a program recorded off the air automatically and then can view it at any time. One can blip out commercials and stop, reverse, or fast forward programs to access any part of the tape, just as one does with a book or magazine. New picture-in-picture systems even allow the viewer to watch as many as nine programs on the same screen.[1] Programs on videocassettes can also be substituted for those offered by the networks or cable companies. This is no different from what readers do with printed materials, but for video viewing it breaks the hammerlock of predetermined programs at set times that the broadcast and cable networks previously offered the viewer. The point here is not to make too much of this token user control, since it is the content of video

FIGURE 1. The UX-80 by Sharp combines the functions of a facsimile machine, a copier, and a telephone in a single object.

FIGURE 2. The form of this videocassette recorder by NEC is much less significant than its multiple functions although the manufacturer is using the form to evoke mythical associations in the user.

programming that is ultimately liberating or confining, but to illustrate that a more flexible relation to the video medium creates the technological condition whereby a person can exercise a greater choice of content.

Today's product flexibility could not have been foreseen by those design theorists and reformers in the past who emphasized the appearance of objects as the central problem to be addressed. When curators at the Museum of Modern Art, for example, stressed the necessity of an object's form to articulate its function, they were thinking only of functions that could be embodied in a shape. The need to rethink the received issues of product form was noted in the 1950s by Tomás Maldonado, Gui Bonsiepe, and others at the Hochschule für Gestaltung in Ulm, West Germany. Focusing on the communicative aspects of forms, they were probably the first educators to incorporate semiotics and information theory into design training. The Ulm faculty recognized the importance of information in the product-user relationship. The symbol system designed by Maldonado and Bonsiepe for IBM data-processing equipment was a good example, and other projects were initiated to clarify the instructional signs on the control panels of various kinds of electronic equipment. Even though the Ulm designers were dealing with less complex objects than those of today, their work remains important for its early emphasis on communication as central to product design.

Since the Hochschule für Gestaltung closed its doors in 1968, the number of functions that many new products possess has grown by large orders of magnitude, enhanced by the creation of microprocessors and software. In computers, software mediates between the user and the object. Since programmers first began to create software, software design and hardware design have been separate processes for the most part. Product designers tend to concentrate on the shape and ergonomics of a physical object, while programmers design software that embodies complex sets of machine commands and capabilities. Despite the fact that machines and software exist in a symbiotic relationship, educators and manufacturers continue to bifurcate the design process. However, as the capabilities of new programming packages increase, there is all the more need to explore the ways that hardware design and software design relate to each other.

With the wider range and greater refinement of machine tasks, the process of giving commands is central to the user, who has come to expect that the machine will take care of the work once it has been correctly instructed. In many instances, however, the user fails to make use of the product's full capabilities. This is certainly true of numerous software programs whose users do not take the time to explore all their possibilities. The disjuncture between expanded product capabilities and

FIGURE 3. AT&T's electronic typewriter includes aspects of a word processor and a printer.

the user's limited ability to use them creates an unprecedented situation for designers, signaling the need for them to address the problem of encouraging and teaching consumers to make the fullest use of new products.

The increasing number of tasks that products can manage prompt the designer to shift radically from the traditional focus on form to the more flexible relation between the product and the user. The designer can no longer foresee all the ways that complex products will be used and must think in terms of multiple possibilities rather than a limited number of set functions.

Although the concept of communicating product functions with the touch of a button or the shift of a mouse is relatively new, user participation in product definition actually has a considerable history. Objects with interchangeable components, such as hand tools, have a long tradition, but flexibility took on more polemical overtones with the design of furniture that had movable parts and was intended to give the user more choices by challenging the conventional expectation of a static form as the end result of the design process. An early example of such furniture was the "Living Structure," a frame with movable platforms that could be adjusted according to user needs, designed by Ken Isaacs, who began developing this type of object in the 1940s. The Living Structure was followed by other designs, such as Joe Colombo's "Additional System," a set of modular units that could be arranged into different combinations of chairs and couches. On a larger scale, the flexible "office landscape" design of the West German Quickborner Team subsequently became popular, as did the "Action Office" produced by the Herman Miller Company under the leadership of designer Robert Probst.

But the user relation to these objects and systems remained a physical one, and the possibilities of exercising their flexibility were self-evident. The connection to products whose functions are not physically defined is different: The individual must become familiar with a lexicon of commands and functions that are either displayed visually on a screen, programmed through a keyboard, or activated by pushing buttons. Such participation is part and parcel of every complex object relationship. And as artificial intelligence and sophisticated sensing devices are incorporated to a greater degree in machine and software design to bring about more sophisticated human-machine interactivity, user participation will become an essential aspect of new products.

Ergonomists have traditionally concentrated on the physical interface between users and objects, but a great need has emerged to learn more about the user's cognitive and emotional relation to products. Whereas designers like Isaacs, Colombo, and Probst thought of the user as someone who manipulated concrete objects, designers and manufac-

FIGURE 4. The Panasonic RC-TX91 can collect and forward phone messages to any number according to preset, timed instructions. It also displays the date and time of each message, can be reprogrammed from another telephone, and includes an automatic sixteen-station dialer.

FIGURE 5. More complex dashboards with digital display screens are projected as standard fare in cars of the future, although it is hard to imagine drivers relating to such a bewildering array of controls while in motion.

turers now are beginning to understand the user as a person who must relate to the functions of a product through a process of information exchange. Operating instructions, whether separate documents or inherent components of the software programs themselves, are central to product use, but user frustration is still all too frequent to take this aspect of design for granted. Yet the process by which a person establishes a relation to a product is an important part of an even larger territory of design, that which I call the product environment. Few designers are prepared to understand this environment, although its importance is likely to grow in years to come.

THE PRODUCT ENVIRONMENT

All the necessary conditions for acquiring the product, learning to use it, following its changes and improvements, providing components for it, and keeping it in good repair are part of the product environment.[2] The term *environment* denotes everything that surrounds the product and becomes part of its identity and value. For chairs and teapots, which have limited functions and little or no maintenance, the environment is a narrow one, but for a personal computer it is wider and embraces many more elements, including the availability of peripherals and software, detailed programming manuals, telephone support systems, electronic bulletin boards, and repair facilities. The computer's environment is the set of elements that satisfies all the conditions for a satisfactory "living" relationship to it. This recognition that a physical object exists in an environment that contributes to its value may be compared to our understanding that our own well-being depends on a healthy natural environment.

For too long, designers, manufacturers, and critics have given inadequate attention to the product environment as a whole. But the design of the physical form is only one element within this environment, the parts of which must all work effectively to achieve the value of good or bad product design. Selected aspects of the product's environment should not merit good design awards if the environment as a whole is deficient in specific aspects.

For the user, the product itself and everything required to make it work are all of a piece. One element essential to high overall quality is efficient maintenance that can be provided without undue stress, either in terms of time or money. A cheap portable audiocassette recorder or an inexpensive telephone answering machine has a deficient product environment: It cannot be repaired for less than the cost of a new one. When something small goes wrong, the purchase price is wasted. As an antidote, Abraham Moles has proposed the concept of the "comprehensive guarantee," something like an insurance contract that would bind the service provider, most likely the manufacturer, to do whatever is necessary to maintain an object's functions during a given period, even replacing a defective product if necessary. Moles perceptively points out that, in fact, people are not so much purchasing an object as a given set of functions on which they depend and which they need to maintain in the most efficient way possible.[3]

Too little attention has been paid to the design of product environments, even though the reasons to redress such neglect are many. Users depend more and more on product efficiency for their daily routines. Often even a simple backup system is lacking, and a key machine out of commission can provoke a crisis. Certain objects have what Ivan Illich calls a "radical monopoly," meaning they are expected to be possessed by everyone. For example, few individuals would try to get by without a telephone and, in some circles, without a telephone answering machine. Likewise, many people would not be able to get to work without an automobile because public transportation systems do not often provide adequate backup for commuters. Thus, manufacturers cannot address every contingency related to satisfactory product use. They need

FIGURE 6. Ameritech promotes cellular telephones by claiming that business executives will be at a disadvantage without them.

to anticipate as many as possible and think in terms of designing a product environment rather than a single object, in order to ensure user satisfaction.

THE NEW USER

Because of the fast pace of life and the constant press of obligations that people in industrialized societies experience, the increased functional capacity of many products has created rising expectations for more and better machine-provided services. Once people know what a machine can offer, they begin to imagine other functions that a machine with additional or expanded capabilities could provide. Encouraged by advertising rhetoric to know no bounds in their expectations, they push manufacturers to design products that are responsive to demands that are ever more refined.

Users expect products to function well because they have little time or money to budget for repairs or for the creation of backup systems. A writer who depends on a computer has a hard time when the machine is down because the files are not accessible—unless a hard copy has been produced as backup. A commuter who needs an automobile every day requires the means for repair as well as the ability to borrow or rent another car. Therefore, the efficient performance of an automobile, backed up by a strong service support system, has become a major selling point for the buyer, while styling, at least in the range of small cars, plays a lesser role than it once did.

People must choose from an immense number of comparable products, as well as a wide range of functions offered by these products. So many options require them to have a clearer sense of their needs to ensure that their time and money are spent efficiently to obtain the payoffs they expect. The danger remains that the user will perceive a new product as useful simply because it offers more efficient functioning, unaware that its acquisition will bring with it a set of obligations regarding service contracts, increased energy use, the supply of parts, and other factors not present with a product that requires no support system.

THE FUTURE

Even though people find themselves in a web of relationships with objects that grow more complicated by the day, manufacturing is nonetheless in a transitional phase between the complex objects of today and the smart, more self-sufficient products of the future. Advances in artificial intelligence and sensor technology have demonstrated that machines can begin to approximate some human perceptual functions and change their operation in relation to such environmental factors as temperature, light, and sound. This technology is already leading to new types of designs, such as the "Smart House," in which more advanced utility systems will combine electrical and communication systems, for example, and will monitor and control energy flows throughout the house more effectively than the average home owner can do now.[4] Sensors in other objects will make them more efficient and accessible. Even as new technology generates products that are friendlier to the user, however, their complexity increases the user's dependence on a specific product environment, which includes responsibility for proper maintenance. While such products enhance user autonomy by performing an expanded range of tasks and providing greater capabilities of communication and production, they also link the individual to an environment that is necessary to keep the product-user relationship in good order.

There is much that designers and manufacturers need to learn about the evolving relationship between products and users. In anticipation of satisfying a need, the user is actually purchasing an environment that promises a satisfying relationship to the product. It is the total organization of such environments that constitutes the next threshold for design. Only by recognizing this can manufacturers, designers, and the public shift their critical faculties to address the issue of product value more accurately.

Notes

[1]I have described only the capabilities of commercially available products. Researchers at the Massachusetts Institute of Technology's Media Lab and elsewhere are experimenting with "intelligent television," which will significantly enhance the flexibility of program formats.

[2]The need to shift from designing objects to designing environments was first suggested by the French sociologist Abraham Moles and was the impetus for the concept put forth here. See Abraham Moles, "The Comprehensive Guarantee: A New Consumer Value," *Design Issues*, 2 (Spring 1985), 53–64, and his chapter in this book.

[3]See Moles, "The Comprehensive Guarantee: A New Consumer Value," p. 9.

[4]See Ralph Lee Smith, *Smart House: The Coming Revolution in Housing* (Columbia, Md.: GP Publishing, 1988).

The Science of Design: Creating the Artificial

Herbert A. Simon

Historically and traditionally, it has been the task of the science disciplines to teach about natural things: how they are and how they work. It has been the task of engineering schools to teach about artificial things: how to make artifacts that have desired properties and how to design.

Engineers are not the only professional designers. Everyone designs who devises courses of action aimed at changing existing situations into preferred ones. The intellectual activity that produces material artifacts is no different fundamentally from the one that prescribes remedies for a sick patient or the one that devises a new sales plan for a company or a social welfare policy for a state. Design, so construed, is the core of all professional training: It is the principal mark that distinguishes the professions from the sciences. Schools of engineering, as well as schools of architecture, business, education, law, and medicine, are all centrally concerned with the process of design.

We gladly acknowledge Professor Simon and MIT Press for granting the permission to reprint these pages.

In view of the key role of design in professional activity, it is ironic that in this century the natural sciences have almost driven the sciences of the artificial from professional school curricula. Engineering schools have become schools of physics and mathematics; medical schools have become schools of biological science; business schools have become schools of finite mathematics. The use of adjectives like "applied" conceals, but does not change, the fact. It simply means that in the professional schools those topics are selected from mathematics and the natural sciences for emphasis which are thought to be most nearly relevant to professional practice. It does not mean that design is taught, as distinguished from analysis.

The movement toward natural science and away from the sciences of the artificial has proceeded further and faster in engineering, business, and medicine than in the other professional fields I have mentioned, though it has by no means been absent from schools of law, journalism, and library science. The stronger universities are more deeply affected than the weaker, and the graduate programs more than the undergraduate. Few doctoral dissertations in first-rate professional schools today deal with genuine design problems, as distinguished from problems in solid-state physics or stochastic processes. I have to make partial exceptions—for reasons I shall mention—of dissertations in computer science and management science, and there are undoubtedly some others, for example, in chemical engineering.

Such a universal phenomenon must have a basic cause. It does have a very obvious one. As professional schools, including the independent engineering schools, are more and more absorbed into the general culture of the university, they hanker after academic respectability. In terms of the prevailing norms, academic respectability calls for subject matter that is intellectually tough, analytic, formalizable, and teachable. In the past much, if not most, of what we knew about design and about the artificial sciences was intellectually soft, intuitive, informal, and cookbooky. Why would anyone in a university stoop to teach or learn about designing machines or planning market strategies when he could concern himself with solid-state physics? The answer has been clear: he usually wouldn't.

The problem is widely recognized in engineering and medicine today and to a lesser extent in business. Some do not think it a problem, because they regard schools of applied science as a superior alternative to the trade schools of the past. If that were the choice, we could agree.[1] But neither alternative is satisfactory. The older kind of professional school did not know how to educate for professional design at an intellectual level appropriate to a university; the newer kind of school has nearly abdicated responsibility for training in the core professional skill.

Thus we are faced with a problem of devising a professional school that can attain two objectives simultaneously: education in both artificial and natural science at a high intellectual level. This too is a problem of design—organizational design.

The kernel of the problem lies in the phrase "artificial science." In my previous chapters I have shown that a science of artificial phenomena is always in imminent danger of dissolving and vanishing. The peculiar properties of the artifact lie on the thin interface between the natural laws within it and the natural laws without. What can we say about it? What is there to study besides the boundary sciences—those that govern the means and the task environment?

The artificial world is centered precisely on this interface between the inner and outer environments; it is concerned with attaining goals by adapting the former to the latter. The proper study of those who are concerned with the artificial is the way in which that adaptation of means to environments is brought about—and central to that is the process of design itself. The professional schools will reassume their professional responsibilities just to the degree that they can discover a science of design, a body of intellectually tough, analytic, partly formalizable, partly empirical, teachable doctrine about the design process.

It is the thesis of this chapter that such a science of design not only is possible but is actually emerging at the present time. It has already begun to penetrate the engineering schools, particularly through programs in computer science and "systems engineering," and business schools through management science. Perhaps it also has beach-heads in other professional curricula, but these are the two with which I am most familiar. We can already see enough of its shape to predict some of the important ways in which engineering schools tomorrow will differ from departments of physics, and business schools from departments of economics and psychology. Let me now turn from questions of university organization to the substance of the matter.

THE LOGIC OF DESIGN: FIXED ALTERNATIVES

We must start with some question of logic.[2] The natural sciences are concerned with how things are. Ordinary systems of logic—the standard propositional and predicate calculi, say—serve these sciences well. Since the concern of standard logic is with declarative statements, it is well suited for assertions about the world and for inferences from those assertions.

Design, on the other hand, is concerned with how things ought to be, with devising artifacts to attain goals. We might question whether the forms of reasoning that are appropriate to natural science are suitable also for design. One might well suppose that introduction of the verb "should" may require additional rules of inference, or modification of the rules already embedded in declarative logic.

Paradoxes of Imperative Logic

Various "paradoxes" have been constructed to demonstrate the need for a distinct logic of imperatives, or a normative, deontic logic. In ordinary logic from "Dogs are pets" and "Cats are pets," one can infer "Dogs and cats are pets." But from "Dogs are pets," "Cats are pets," and "You should keep pets," one can infer "You should keep cats and dogs?" And from "Give me needle and thread!" can one deduce, in analogy with declarative logic, "Give me needle or thread!"? Easily frustrated people would perhaps rather have neither needle nor thread than one without the other, and peaceloving people, neither cats nor dogs, rather than both.

As a response to these challenges of apparent paradox, there have been developed a number of constructions of modal logic for handling "should," "shalts," and "oughts" of various kinds. I think it is fair to say that none of these systems has been sufficiently developed or sufficiently widely applied to demonstrate that it is adequate to handle the logical requirements of the process of design.

Fortunately, such a demonstration is really not essential, for it can be shown that the requirements of design can be met fully by a modest adaptation of ordinary declarative logic. Thus a special logic of imperatives is unnecessary.[3]

Reduction to Declarative Logic

The easiest way to discover what kinds of logic are needed for design is to examine what kinds of logic designers use when they are being careful about their reasoning. Now there would be no point in doing this if designers were always sloppy fellows who reasoned loosely, vaguely, and intuitively. Then we might say that whatever logic they used was not the logic they *should* use.

However, there exists a considerable area of design practice where standards of rigor in inference are as high as one could wish. I refer to the domain of so-called "optimization methods," most highly developed in statistical decision theory and management science but acquiring growing importance also in engineering design theory. The theories of probability and utility, and their intersection, have received the painstaking attention not only of practical designers and decision makers but also of a considerable number of the most distinguished logicians and mathematicians of the present and recent past generations. F. P. Ramsey, B. de Finetti, A. Wald, J. von Neumann, J. Neyman, K. Arrow, and L. J. Savage are examples.

The logic of optimization methods can be sketched as follows: The "inner environment" of the design problem is represented by a set of given alternatives of action. The alternatives may be given *in extenso*; more commonly they are specified in terms of *command variables* that have defined domains. The "outer environment" is represented by a set of parameters, which may be known with certainty or only in terms of a probability distribution. The goals for adaptation of inner to outer environment are defined by a utility function—a function, usually scalar, of the command variables and environmental parameters—perhaps supplemented by a number of constraints (inequalities, say, between functions of the command variables and environmental parameters). The optimization problem is to find an admissible set of values of the command variables, compatible with the constraints, that maximize the utility function for the given values of the environmental parameters. (In the probabilistic case we might say, "maximize the expected value of the utility function," for instance, instead of "maximize the utility function.")

A stock application of this paradigm is the so-called "diet problem" shown in Figure 1. A list of foods is provided, the command variables being quantities of the various foods to be included in the diet. The environmental parameters are the prices and nutritional contents (calories, vitamins, minerals, and so on) of each of the foods. The utility function is the cost (with a minus sign attached) of the diet, subject to the constraints, say, that it not contain more than 2,000 calories per day, that it meet specified minimum needs for vitamins and minerals, and that rutabaga not be eaten more than once a week. The constraints may be viewed as characterizing the inner environment. The problem is to select the quantities of foods that will meet the nutritional requirements and side conditions at the given prices for the lowest cost.

The diet problem is a simple example of a class of problems that are readily handled, even when the number of variables is exceedingly large, by the mathematical formalism known as linear programming. I shall come back to the technique a little later. My present concern is with the logic of the matter.

Since the optimization problem, once formalized, is a standard mathematical problem—to maximize a function subject to constraints—it is evident that the logic used to deduce the answer is the standard logic of the predicate calculus on which mathematics rests. How does the formalism avoid making use of a special logic of imperatives? It does so by dealing with sets of *possible worlds*: First consider all the possible worlds that meet the constraints of the outer environment; then find

Logical Terms		*Example: The diet problem*
Command variables	("Means")	Quantities of foods
Fixed parameters	("Laws")	Prices of foods Nutritional contents
Constraints	("Ends")	Nutritional requirements
Utility function		Cost of diet

Constraints characterize the inner environment; parameters characterize the outer environment.

Problem: Given the constraints and fixed parameters, find values of the command variables that maximize utility.

The paradigm for imperative logic

the particular world in the set that meets the remaining constraints of the goal and maximizes the utility function. The logic is exactly the same as if we were to adjoin the goal constraints and the maximization requirement, as new "natural laws," to the existing natural laws embodied in the environmental conditions.[4] We simply ask what values the command variables *would* have in a world meeting all these conditions and conclude that these are the values the command variables *should* have.

Computing the Optimum

Our discussion thus far has already provided us with two central topics for the curriculum in the science of design:

1. *Utility theory and statistical decision theory as a logical framework for rational choice among given alternatives.*
2. *The body of techniques for actually deducing which of the available alternatives is the optimum.*

Only in trivial cases is the computation of the optimum alternative an easy matter. If utility theory is to have application to real-life design problems, it must be accompanied by tools for actually making the computations. The dilemma of the rational chess player is familiar to all. The optimal strategy in chess is easily demonstrated: Simply assign a value of +1 to a win, 0 to a draw, and −1 to a loss; consider all possible courses of play; minimax backward from the outcome of each, assuming each player will take the most favorable move at any given point. This procedure will determine what move to make now. The only trouble is that the computations required are astronomical (the number 10 is often mentioned in this context) and hence cannot be carried out—not by humans, not by existing computers, not by prospective computers.

A theory of design as applied to the game of chess would encompass not only the utopian minimax principle but also some practicable procedures for finding good moves in actual board positions in real time, within the computational capacities of real human beings or real computers. No exceptionally good procedures of this kind exist today, other than those stored in the memories of grandmasters, but there is at least one computer program that plays at the level of an expert or a weak master—that is, better than all save a few hundred human players.

The second topic then for the curriculum in the science of design consists in the efficient computational techniques that are available for actually finding optimum courses of action in real situations, or reasonable approximations to real situations. That topic has a number of important components today, most of them developed—at least to the level of practical application—within the past 25 years. These include linear programming theory, dynamic programming, geometric programming, queuing theory, and control theory.

THE LOGIC OF DESIGN; FINDING ALTERNATIVES

When we take up the case where the design alternatives are not given in any constructive sense but must be synthesized, we must ask once more whether any new forms of reasoning are involved in the synthesis, or whether again the standard logic of declarative statements is all we need.

In the case of optimization, we asked: "Of all possible worlds (those attainable for some admissible values of the action variables), which is the best (yields the highest value of the criterion function)?" As we saw, this is a purely empirical question, calling only for facts and ordinary declarative reasoning to answer it.

In this case, where we are seeking a satisfactory alternative, once we have found a candidate we can ask: "Does this alternative satisfy all the design criteria?" Clearly this is also a factual question and raises no new issues of logic. But how about the process of *searching* for candidates? What kind of logic is needed for the search?

Means-Ends Analysis

The condition of any goal-seeking system is that it is connected to the outside environment through two kinds of channels: the afferent, or sensory, channels through which it receives information about the environment and the efferent, or motor, channels through which it acts on the environment.[5] The system must have some means of storing in its memory information about states of the world—afferent, or sensory, information—and information about actions—efferent, or motor, information. Ability to attain goals depends on building up associations, which may be simple or very complex, between particular changes in states of the world and particular actions that will (reliably or not) bring these changes about. We have described these associations as productions.

Except for a few built-in reflexes, an infant has no basis for correlating his sensory information with his actions. A very important part of his early learning is that particular actions or sequences of actions will bring about particular changes in the state of the world as he senses it. Until he builds up this knowledge, the world of sense and the motor world are two entirely separate, entirely unrelated worlds. Only as he begins to acquire experience as to how elements of the one relate to elements of the other can he act purposefully on the world.

The computer problem-solving program called GPS, designed to model some of the main features of human problem solving, exhibits in stark form how goal-directed action depends on building this kind of bridge between the afferent and the efferent worlds. On the afferent, or sensory, side, GPS must be able to represent desired situations or desired objects as well as the present situation. It must be able also to represent *differences* between the desired and the present. On the efferent side, GPS must be able to represent *actions* that change objects or situations. To behave purposefully, GPS must be able to select from time to time those particular actions that are likely to remove the particular differences between desired and present states that the system detects. In the machinery of GPS, this selection is achieved through a *table of connections*, which associates with each kind of detectable difference those actions that are relevant to reducing that difference. These are its associations, in the form of productions, which relate the afferent to the efferent world. Since reaching a goal generally requires a sequence of actions, and since some attempts may be ineffective, GPS must also have means for detecting the progress it is making (the changes in the differences between the actual and the desired) and for trying alternate paths.

Now the real worlds to which problem solvers and designers address themselves are seldom completely additive in this sense. Actions have side consequences (may create new differences) and sometimes can only be taken when certain side conditions are satisfied (call for removal of other differences before they become applicable). Under these circumstances one can never be certain that a partial sequence of actions that accomplishes *certain* goals can be augmented to provide a solution that satisfied *all* the conditions and attains *all* the goals (even though they be satisficing goals) of the problem.

For this reason problem-solving systems and design procedures in the real world do not merely *assemble* problem solutions from components but must *search* for appropriate assemblies. In carrying out such a search, it is often efficient to divide one's eggs among a number of baskets—that is, not to follow out one line until it succeeds completely or fails definitely but to begin to explore several tentative paths, continuing to pursue a few that look most promising at a given moment. If one of the active paths begins to look less promising, it may be replaced by another that had previously been assigned a lower priority.

Our discussion of design when the alternatives are not given has yielded at least three additional topics for instruction in the science of design:

3. *Adaptation of standard logic to the search for alternatives.* Design solutions are sequences of actions that lead to possible worlds satisfying specified constraints. With satisficing goals the sought-for possible worlds are seldom unique; the search is for *sufficient*, not *necessary*, actions for attaining goals.
4. *The exploitation of parallel, or near-parallel, factorizations of differences.* Means-ends analysis is an example of a broadly applicable problem-solving technique that exploits this factorization.
5. *The allocation of resources for search to alternative, partly explored action sequences.* I should like to elaborate somewhat on this last-mentioned topic.

DESIGN AS RESOURCE ALLOCATION

There are two ways in which design processes are concerned with the allocation of resources. First, conservation of scarce resources may be one of the criteria for a satisfactory design. Second, the design process itself involves management of the resources of the designer, so that his efforts will not be dissipated unnecessarily in following lines of inquiry that prove fruitless.

There is nothing special that needs to be said here about resource conservation—cost minimization, for example, as a design criterion. Cost minimization has always been an implicit consideration in the design of engineering structures, but until a few years ago it generally *was* only implicit, rather than explicit. More and more cost calculations have been brought explicitly into the design procedure, and a strong case can be made today for training design engineers in that body of technique and theory that economists know as "cost-benefit analysis."

THE SHAPE OF THE DESIGN: HIERARCHY

I have given some reasons why complex systems might be expected to be constructed in a hierarchy of levels, or in a boxes-within-boxes form. The basic idea is that the several components in any complex system will perform particular subfunctions that contribute to the over-all function. Just as the "inner environment" of the whole system may be defined by describing its functions, without detailed specification of its mechanisms, so the "inner environment" of each of the subsystems may be defined by describing the functions of that subsystem, without detailed specification of *its* submechanisms.[6]

To design such a complex structure, one powerful technique is to discover viable ways of decomposing it into semi-independent components corresponding to its many functional parts. The design of each component can then be carried out with some degree of independence of the design of others, since each will affect the others largely through its function and independently of the details of the mechanisms that accomplish the function.[7]

There is no reason to expect that the decomposition of the complete design into functional components will be unique. In important instances there may exist alternative feasible decompositions of radically different kinds. This possibility is well known to designers of administrative organizations, where work can be divided up by subfunctions, by subprocesses, by subareas, and in other ways. Much of classical organization theory in fact was concerned precisely with this issue of alternative decompositions of a collection of interrelated tasks.

The Generator-Test Cycle

One way of considering the decomposition, but acknowledging that the interrelations among the components cannot be ignored completely, is to think of the design process as involving, first, the generation of alternatives and, then, the testing of these alternatives against a whole array of requirements and constraints. There need not be merely a single generate-test cycle, but there can be a whole nested series of such cycles. The generators implicitly define the decomposition of the design problem, and the tests guarantee that important indirect consequences will be noticed and weighed. Alternative decompositions correspond to different ways of dividing the responsibilities for the final design between generators and tests.

To take a greatly oversimplified example, a series of generators may generate one or more possible outlines and schemes of fenestration for a building, while tests may be applied to determine whether needs for particular kinds of rooms can be met within the outlines generated. Alternatively the generators may be used to evolve the structure of rooms, while tests are applied to see whether they are consistent with an acceptable over-all shape and design. The house can be designed from the outside in or from the inside out.[8]

Alternatives are also open, in organizing the design process, as to how far development of possible subsystems will be carried before the over-all coordinating design is developed in detail, or vice-versa—how far the over-all design should be carried before various components, or possible components, are developed. These alternatives of design are familiar to architects. They are familiar also to composers, who must decide how far the architectonics of a musical structure will be evolved before some of the component musical themes and other elements have been invented. Computer programmers face the same choices, between working downward from executive routines to subroutines or upward from component subroutines to a coordinating executive. A theory of design will include principles—most of which do not yet exist—for deciding such questions of precedence and sequence in the design process.

Process as a Determinant of Style

When we recall that the process will generally be concerned with finding a satisfactory design, rather than an optimum design, we see that sequence and the division of labor between generators and tests can affect not only the efficiency with which resources for designing are used but also the nature of the final design as well. What we ordinarily call "style" may stem just as much from these decisions about the design process as from alternative emphases on the goals to be realized through the final design. An architect who designs buildings from the outside in will arrive at quite different buildings from one who designs from the inside out, even though both of them might agree on the characteristics that a satisfactory building should possess.

When we come to the design of systems as complex as cities, or buildings, or economies, we must give up the aim of creating systems that will optimize some hypothesized utility function, and we must consider whether differences in style of the sort I have just been describing do not represent highly desirable variants in the design process rather than alternatives to be evaluated as "better" or "worse." Variety, within the limits of satisfactory constraints, may be a desirable end in itself, among other reasons, because it permits us to attach value to the search as well as its outcome—to regard the design process as itself a valued activity for those who participate in it.

We have usually thought of city planning as a means whereby the planner's creative activity could build a system that would satisfy the needs of a populace. Perhaps we should think of city planning as a valuable creative activity in which many members of a community can have the opportunity of participating—if we have wits to organize the process that way.

However that may be, I hope I have illustrated sufficiently that both the shape of the design and the shape and organization of the design process are essential components of a theory of design. These topics constitute the sixth item in my proposed curriculum in design:

6. *The organization of complex structures and its implication for the organization of design processes.*

REPRESENTATION OF THE DESIGN

I have by no means surveyed all facets of the emerging science of design. In particular I have said little about the influence of problem representation on design. Although the importance of the question is recognized today, we have little systematic knowledge about it. I shall cite one example, to make clear what I mean by "representation."

Here are the rules of a game, which I shall call number scrabble. The game is played by two people with nine cards—let us say the ace through the nine of hearts. The cards are placed in a row, face up, between the two players. The players draw alternately, one at a time, selecting any one of the cards that remain in the center. The aim of the game is for a player to make up a "book," that is, a set of exactly three cards whose spots add to 15, before his opponent can do so. The first player who makes a book wins; if all nine cards have been drawn without either player making a book, the game is a draw.

What is a good strategy in this game? How would you go about finding one? If the reader has not already discovered it for himself, let me show how a change in representation will make it easy to play the game well. The magic square here is made up of the numerals from 1 through 9.

Each row, column, or diagonal adds to 15, and every triple of these numerals that add to 15 is a row, column, or diagonal of the magic square. From this, it is obvious that "making a book" in number scrabble

4	9	2
3	5	7
8	1	6

is equivalent to getting "three in a row" in a game of tic-tac-toe. But most people know how to play tic-tac-toe well, hence they can simply transfer their usual strategy to number scrabble.[9]

Problem Solving as Change in Representation

That representation makes a difference is a long-familiar point. We all believe that arithmetic has become easier since Arabic numerals and place notation replaced Roman numerals, although I know of no theoretical treatment that explains why.[10] That representation makes a difference is evident for a different reason. All mathematics exhibits in its conclusions only what is already implicit in its premises, as I mentioned in a previous chapter. Hence all mathematical derivation can be viewed simply as change in representation, making evident what was previously true but obscure.

This view can be extended to all of problem solving. Solving a problem simply means representing it so as to make the solution transparent.[11] If the problem solving could actually be organized in these terms, the issue of representation would indeed become central. But even if it cannot—if this is too exaggerated a view—a deeper understanding of how representations are created and how they contribute to the solution of problems will become an essential component in the future theory of design.

Spatial Representation

Since much of design, particularly architectural and engineering design, is concerned with objects or arrangements in real Euclidean two-dimensional or three-dimensional space, the representation of space and of things in space will necessarily be a central topic in a science of design. From our previous discussion of visual perception, it should be clear that "space" inside the head of the designer of the memory of a computer may have very different properties from a picture on a paper or a three-dimensional model.

These representational issues have already attracted the attention of those concerned with computer-aided design—the cooperation of human and computer in the design process. As a single example, I may mention Ivan Sutherland's SKETCHPAD program, which allows geometric shapes in terms of constraints, to which they then conform.[12]

Geometric considerations are also prominent in the attempts to automate completely the design, say, of printed or etched circuits, or of buildings. Grason, for example, in a system for designing house floor plans, constructs an internal representation of the layout that helps one decide whether a proposed set of connections among rooms, selected to meet design criteria for communication, and so on, can be realized in a plane.[13]

The Taxonomy of Representation

An early step toward understanding any set of phenomena is to learn what kinds of things there are in the set—to develop a taxonomy. This step has not yet been taken with respect to representations. We have only a sketchy and incomplete knowledge of the different ways in which problems can be represented and much less knowledge of the significance of the differences.

In a completely pragmatic vein, we know that problems can be described verbally, in natural language. They often can be described mathematically, using standard formalisms of algebra, geometry, set theory, analysis, or topology. If the problems relate to physical objects, they (or their solutions) can be represented by floor plans, engineering drawings, renderings, or three-dimensional models. Problems that have to do with actions can be attacked with flow charts and programs.

Other items will most likely need to be added to the list, and there may exist more fundamental and significant ways of classifying its members. But even though our classification is incomplete, and perhaps superficial, we can begin to build a theory of the properties of these representations. A number of topics in the growing theories of machines and of programming languages may give us some notion of the directions that a theory of representations—at least on its more formal side—may take.[14] These topics can also provide, at the beginning, some of the substance for the final subject in our program on the theory of design:

7. *Alternative representations for design problems.*

SUMMARY—TOPICS IN THE THEORY OF DESIGN

My main goal in this chapter has been to show that there already exist today a number of components of a theory of design and a substantial body of knowledge, theoretical and empirical, relating to each. As we draw up our curriculum in design—in the science of the artificial—to take its place by the side of natural science in the whole engineering curriculum, it includes at least the following topics:

The evaluation of designs

1. Theory of evaluation: utility theory, statistical decision theory
2. Computational methods:
   a. Algorithms for choosing *optimal* alternatives, such as linear programming computations, control theory, dynamic programming
   b. Algorithms and heuristics for choosing *satisfactory* alternatives
3. The formal logic of design: imperative and declarative logics

The search for alternatives

4. Heuristic search: factorization and means-ends analysis
5. Allocation of resources for search
6. Theory of structure and design organization: hierarchic systems
7. Representation of design problems

In small segments of the curriculum—the theory of evaluation, for example, and the formal logic of design—it is already possible to organize the instruction within a framework of systematic, formal theory. In many other segments the treatment would be more pragmatic, more empirical.

But nowhere do we need to return or retreat to the methods of the cookbook that originally put design into disrepute and drove it from the engineering curriculum. For there exist today a considerable number of examples of actual design processes, of many different kinds, that have been defined fully and cast in the metal, so to speak, in the form of running computer programs: optimizing algorithms, search procedures, and special-purpose programs for designing motors, balancing assembly lines, selecting investment portfolios, locating warehouses, designing highways, diagnosing and treating diseases, and so forth.

Because these computer programs describe complex design processes in complete, painstaking detail, they are open to full inspection and analysis, or to trial by simulation. They constitute a body of empirical phenomena to which the student of design can address himself and which he can seek to understand. There is no question, since these programs exist, of the design process hiding behind the cloak of "judgement" or "experience." Whatever judgment or experience was used in creating the programs must now be incorporated in them and hence be observable. The programs are the tangible record of the variety of schemes that man has devised to explore his complex outer environment and to discover in that environment the paths to his goals.

ROLE OF DESIGN IN THE LIFE OF THE MIND

I have called my topic "the theory of design" and my curriculum a "program in design." I have emphasized its role as a complement to the natural science curriculum in the total training of a professional engineer—or of any professional whose task is to solve problems, to choose, to synthesize, to decide.

But there is another way in which the theory of design may be viewed in relation to other knowledge. I have discussed psychology—specifically on man's relation to his biological inner environment. The present chapter may also be construed as a chapter on psychology: on man's relation to the complex outer environment in which he seeks to survive and achieve.

All of this, so construed, has import that goes beyond the professional work of the man we have called the "designer." Many of us have been unhappy about the fragmentation of our society into two cultures. Some of us even think there are not just two cultures but a large number of cultures. If we regret that fragmentation, then we must look for a common core of knowledge that can be shared by the members of all cultures—a core that includes more significant topics than the weather, sports, automobiles, the care and feeding of children, or perhaps even politics. A common understanding of our relation to the inner and outer environments that define the space in which we live and choose can provide at least part of that significant core.

This may seem an extravagant claim. Let me use the realm of music to illustrate what I mean. Music is one of the most ancient of the sciences of the artificial, and was so recognized by the Greeks. Anything I have said about the artificial would apply as well to music, its composition or its enjoyment, as to the engineering topics I have used for most of my illustrations.

Music involves a formal pattern. It has few (but important) contacts with the inner environment; that is, it is capable of evoking strong emotions, its patterns are detectable by human listeners, and some of its harmonic relations can be given physical and physiological interpretations (though the esthetic import of these is debatable). As for the outer environment, when we view composition as a problem in design, we encounter just the same tasks of evaluation, of search for alternatives, and of representation that we do in any other design problem. If it pleases us, we can even apply to music some of the same techniques of automatic design by computer that have been used in other fields of design. If computer-composed music has not yet reached notable heights of esthetic excellence, it deserves, and has already received, serious attention from professional composers and analysts, who do not find it written in tongues alien to them.[15]

Undoubtedly there are tone-deaf engineers, just as there are mathematically ignorant composers. Few engineers and composers, whether deaf, ignorant, or not, can carry on a mutually rewarding conversation about the content of each other's professional work. What I am suggesting is that they *can* carry on such a conversation about design, can begin to perceive the common creative activity in which they are both engaged, can begin to share their experiences of the creative, professional design process.

Those of us who have lived close to the development of the modern computer through gestation and infancy have been drawn from a wide variety of professional fields, music being one of them. We have noticed the growing communication among intellectual disciplines that takes place around the computer. We have welcomed it, because it has brought us into contact with new worlds of knowledge—has helped us combat our own multiple-cultures isolation. This breakdown of old disciplinary boundaries has been much commented upon and its connection with computers and the information sciences often noted.

But surely the computer, as a piece of hardware, or even as a piece of programmed software, has nothing to do directly with the matter. I have already suggested a different explanation. The ability to communicate across fields—the common ground—comes from the fact that all who use computers in complex ways are using computers to design or to participate in the process of design. Consequently we as designers, or as designers of design processes, have had to be explicit as never before about what is involved in creating a design and what takes place while the creation is going on.

The real subjects of the new intellectual free trade among the many cultures are our own thought processes, our processes of judging, deciding, choosing, and creating. We are importing and exporting from one intellectual discipline to another ideas about how a serially organized information-processing system like a human being—or a computer, or a complex of men and women and computers in organized cooperation—solves problems and achieves goals in outer environments of great complexity.

The proper study of mankind has been said to be man. But I have argued that man—or at least the intellective component of man—may be relatively simple, that most of the complexity of his behavior may be drawn from man's environment, from man's search for good designs. If I have made my case, then we can conclude that, in large part, the proper study of mankind is the science of design, not only as the professional component of a technical education but as a core discipline for every liberally educated person.

Notes

[1]That was in fact the choice in our engineering schools a generation ago. The schools needed to be purged of vocationalism, and a genuine science of design did not exist even in a rudimentary form as an alternative. Hence the road forward was the road toward introducing more fundamental science. Karl Taylor Compton was one of the prominent leaders in this reform, which was a main theme in his presidential inaugural address at MIT in 1930: "I hope . . . that increasing attention in the Institute may be given to the fundamental sciences; that they may achieve as never before the spirit and results of research; that all courses of instruction may be examined carefully to see where training in details has been unduly emphasized at the expense of the more powerful training in all-embracing fundamental principles." Notice that President Compton's emphasis was on "fundamental," an emphasis as sound today as it was in 1930. What I am urging in this essay is not a departure from the fundamental but an inclusion in the curriculum of the fundamental in engineering along with the fundamental in natural science. That was not possible in 1930, but it is possible today.

[2]I have treated the question of logical formalism for design at greater length in two earlier papers: "The Logic of Rational Decision," *British Journal for the Philosophy of Science*, 16 (1965), 169–86 and "The Logic of Heuristic Decision Making," in Nicholas Rescher, ed., *The Logic of Decision and Action* (Pittsburgh, Pa.: University of Pittsburgh Press, 1967), pp. 1–35. The present discussion is based on these two papers, which have been reprinted as Chapters 3.1 and 3.2 in my *Models of Discovery* (Dordrecht: D. Reidel, 1977).

[3]I should like to stress the word *unnecessary*. When I said something like this in another place (the second paper mentioned in the previous footnote), an able logician, who had specialized in modal logics, accused me of asserting that modal logics were "impossible." Now this is patently false: modal logics can be shown to exist in the same way that giraffes can—namely, by exhibiting some of them. The question is not whether they exist but what they are good for. A modal logician should have no difficulty in distinguishing "non-necessity" from "impossibility."

[4]The use of the notion of "possible worlds" to embed the logic of imperatives in declarative logic goes back at least to Jürgen Jürgensen, "Imperatives and Logic," *Erkenntnis*, 7 (1937–1938), 288–96. See also my *Administrative Behavior* (New York: Macmillan, 1947), chapter 3. More recently this same idea has been used by several logicians to construct a formal bridge between the predicate calculus and modal logic by means of so-called semantic or model-theoretic methods. See, for example, Richard Montague, "Logical Necessity, Physical Necessity, Ethics, and Quantifiers," *Inquiry*, 4 (1960), 259–69, where references are also given to work of Stig Kanger and Saul Kripke, and Jaakko Hintikka, "Modality and Quantification," *Theoria*, 27 (1961), 119–28. While these model-theoretic proposals are basically sound, none of them seems yet to have given adequate attention to the special role played in the theory by command variables and criterial constraints.

[5]Notice that we are not saying that the two kinds of channels operate independently of each other, since they surely do not in living organisms, but that we can distinguish conceptually, and to some extent neurologically, between the incoming and outgoing flows.

[6]I have developed this argument at greater length in my essay "The Architecture of Complexity," chapter 7.

[7]This approach to the design of complex structures has been explored by Christopher Alexander in *Notes on the Synthesis of Form* (Cambridge, Mass.: Harvard University Press, 1967). He has also presented in his book some automated procedures for finding plausible decompositions once the matrix of interconnections of component functions has been specified.

[8]I am indebted to John Grason for many ideas on the topic of this section. J. Grason,

"Fundamental Description of a Floor Plan Design Program," EDRA1, *Proceedings of the First Environmental Design Association Conference*, ed. H. Sanoff and S. Cohn, North Carolina State University, 1970.

[9]Number scrabble is not the only isomorph of tic-tac-toe. John A. Michon has described another, JAM, which is the dual of tic-tac-toe in the sense of projective geometry. That is, the rows, columns, and diagonals of tic-tac-toe become points in JAM, and the squares of the former become line segments joining the points. The game is won by "jamming" all the segments through a point—a move consists of seizing or jamming a single segment. Other isomorphs of tic-tac-toe are known as well.

[10]My colleague, Allen Newell, has been investigating this question. I shall not try to anticipate his answer.

[11]Saul Amarel, "On the Mechanization of Creative Processes," IEEE *Spectrum*, 3 (April 1966), 112–14.

[12]I. E. Sutherland, "SKETCHPAD, A Man-Machine Graphical Communication System," *Proceedings, AFIPS Spring Joint Computer Conference, 1963* (Baltimore: Spartan Books, 1963), pp. 329–46.

[13]See also C. E. Pfefferkorn, "The Design Problem Solver: A System for Designing Equipment or Furniture Layouts," in C. M. Eastman, ed., *Spatial Synthesis in Computer-Aided Building Design* (London: Applied Science, 1975).

[14]By way of example, see Marvin L. Minsky, *Computation: Finite and Infinite Machines* (Englewood Cliffs, N.J.: Prentice-Hall, 1967) and Kenneth E. Iverson, *A Programming Language* (New York: Wiley, 1962).

[15]L. A. Hillier and L. M. Isaacson's *Experimental Music* (New York: McGraw-Hill, 1959), reporting experiments begun more than two decades ago, still provides a good introduction to the subject of musical composition viewed as design. See also Walter R. Reitman, *Cognition and Thought* (New York: Wiley, 1965), chapter 6, "Creative Problem Solving: Notes from the Autobiography of a Fugue."

Goals and Problems

Norbert Wiener

The years since the Second World War have seen an enormous change in all our policies, and not least in our scientific policies. It has been a period of great projects, of the partial eclipse of the individual, and of the emergence of a mass attack as a normal method of scientific work. It has seen the universities and other classically scientific organizations and organizations of learning, facing a period of tremendous inflation which has decimated their resources and has made it necessary to address new appeals to the public.

The following text reproduces a chapter, the last, of a complete manuscript for a book entitled "The Philosophy of Invention" that Norbert Wiener was revising until a few weeks before his death. Surprisingly enough, the book was never published, the manuscript almost disappeared, and it was certainly forgotten by all of Wiener's biographers. It is an important contribution that deserves to be rediscovered. (This text was found in 1983 in the archives of MIT while researching materials for an intellectual biography of Norbert Wiener.) We are grateful to Helen W. Samuels, institute archivist and head of special collections, for granting us the permission to publish the manuscript—Marco Diani.)

This public has become emergency-minded and therefore prepared to sacrifice long-time ends and ideals of scholarship to the combating of the dangers of the moment. These dangers include a world struggle for supremacy between the United States and Soviet Russia. Like the United States, Russia is an aggressive nation with a philosophy conditioned not so much upon stability as on continued progress toward a goal not yet attained.

However, our struggle with Soviet Russia and the Soviet complex of nations, which has already dwarfed the indecisive medieval struggle between Christianity and Islam, is only one of those which the present generation must face. Militant capitalism—and not all capitalism is militant—is a creed which represents as much of a departure from much of our nineteenth-century traditions as does Communism, particularly when it is combined with a pseudo-egalitarianism which not only claims that one man is as good as another, in his moral rights and duties, but that it is interchangeable with another in his abilities.

The present disputes and controversies in which we have engaged have lasted over decades. However, chronic they may prove to be, they have attracted our attention and concern in an acute form, which I do not think has yet settled down to a long-time equilibrium. In this phase as short-time conflicts, they have drawn our attention away from our long-time problems. They have even made large sectors of our public opinion scarcely aware that the long-time problems exist. I have asserted, and I think that to a certain degree I have shown, that the long-time problems must be faced by long-time social feedbacks, or, if you prefer, checks and balances, if we are not to sell the future of humanity short.

I have already pointed out that in an athletic contest which demands vigorous exertion and even maximum exertion, of short duration, we can ignore with impunity the long-time needs of the athlete. The sprint runner may burst his heart if he is out of condition, but he will not die of starvation because of the immediate demands of the contest. He will not even die of oxygen starvation, for there is a considerable amount of oxygen already held in combination by the hemoglobin of his blood and of his tissues. He cannot exhaust this reserve in a short sprint.

But not all athletic contests are of short duration. No marathon racer can continue to run if his rate of consumption of oxygen is greater than the rate of intake of oxygen into his blood. He must reach a quasi-equilibrium in his needs which will last for the whole period of the race. The legend of Philippides going mad and dying at the end of his run from Marathon to Athens is exactly what one must expect on such an occasion. Similarly, a nation, or the human race, which must face a protracted period of emergency, can by no means ignore the continual satisfaction of its long-time needs.

Now, a great war does not usually last more than some five years, and even the cold war, which belongs to the times in which we live, has not yet endured more than a decade. When, however, our President tells us that the emergency of the cold war and the threat of a new hot war will be with us for some 40 years, we cannot be content with meeting this threat by measures which depend solely on the consumption of our stored resources.

Forty years from now, our leading inventions are likely to be based on scientific ideas not yet even contemplated. The relative rate at which Russia and ourselves store up reserves of fundamental science is more likely to be a fair measure of our ability to meet Russian competition 40 years from now, than is our present level of technical achievement on the basis of the scientific concepts which we have already mastered. There is a certain and very real evidence that Russia may have been devoting more thought to the long-time problem of maintaining the level of science than we have. If we are not careful, this may offset all our technical know-how and all the concentrated large-scale effort which we have shown ourselves as capable of making.

Russia may be our chief antagonist or potential antagonist for many years, perhaps even for more than the 40 of which President Eisenhower speaks. However, our long-time chief antagonist will not be Russia, but is to be sought for among the continuing threats of hunger, thirst, ignorance, over-population, and perhaps the new dangers of the poisoning of the world in which we live by the radioactive by-products of an atomic age. Thus we must go into training for a marathon race and not for a sprint. We cannot win in this marathon race unless we have that sense of the future which is based on a sense of history.

This book, then, is a plea that even in this present difficult and confusing period, we must take a long view and conserve long-time values. I will not say that our present effort is not enough, for it is nearly a maximum effort. It is certainly not, in its entirety, the right sort of effort for the problems which will continue to confuse us. We cannot survive our long prospective ordeal by virtue of adrenalin alone.

Let me sum up certain considerations already explicit or implicit. We have already seen that one of the basic observations of the history of invention and discovery is that the really great steps in invention are, at least in many cases, nothing but the embodiments of changes in the intellectual climate, which often antedate their industrial use by several decades. In this pre-industrial stage, the new ideas might or might not have been introduced, and the lack of one or two individuals in the chain of thought leading up to them, while it might not have made the new developments forever impossible, could easily have delayed them for over a generation.

On the other hand, the later stages of invention which often result from the building up of general scientific education, and of thought in many separate fields, to the point where the new ideas can be appreciated, has a very strong tendency to be a repeated and not a sporadic phenomenon, and to take place nearly at the same time in many fields of work, among many individuals, and in many countries. What I wish to discuss is how such a multiform human activity can best be allowed to develop in its fullness and what measures suggest themselves for fostering it and its various diverse stages.

By the time that the second period of invention is well along, and the inventors are working with ideas which thrust themselves upon them from many different quarters, it slowly begins to be possible to evaluate the repercussion they will have on human affairs generally, and on manufacturing and engineering technique in particular. The questions of legal ownership and protection still need the expert services of the patent lawyer. Apart from this patent phase of the development of invention, it is possible to envisage a program of development and exploitation which, if it involves elements of risk, on the whole involves a calculable risk.

With such a calculable risk in view, it may easily pay a commercial organization, or a pseudo-commercial organization of the nature now to be found in many government departments, to put onto the idea a large and subdivided group of engineers and scientists, and even to assign each man in that group a specific task which he is to accomplish. If this development were completely without risk, it would probably also be without reward. The picture I am giving contemplates a certain risk, and bears a not-too-remote resemblance to a commercial undertaking of the ordinary sort in which new ideas do not necessarily play an essential role.

The mass-attack, piece-work concept of the development, research, and engineering functions of large organizations has captured both the public fancy and the fancy of our entrepreneurs and government officials. It has captured this fancy almost to the exclusion of any other concept of discovery or invention. I have said and I here say again that it is not a concept which can be extended to the complete history of discovery, and which is particularly inapplicable to the first and most vital step toward invention.

In their early stage, invention and discovery are not calculable risks. In the first place, the really inventive mind must take chances. A long-time record of no false starts probably does not mean infallibility on the part of the inventor, but merely that he has not been willing to push his ideas as far as they merit. The baseball player without a record of errors is the one who does not go after the balls that he might barely pull down, but allows them to damage the record of the other fellow.

In the second place, the really big finds in invention and discovery take a long period to mature. Very often they cannot be put to any economic use until other ideas as yet undiscovered have rendered practical what was once a long shot. There is no way of keeping an invention on ice for a long period and still exploiting it commercially. To keep a discovery secret means an unwilling or willing denial of the idea advantages of the discovery to other people. After a very limited period—limited as far as really fundamental inventions are concerned—to publish it means to dedicate it to the public. As we have seen in the case of Heaviside, this means to deny exclusive commercial rights to any person. Thus the probability that a really fundamental invention will accrue to the commercial advantage of the person or group who has invented it, is very small indeed. It is simply not worthwhile to invest a cent in it, as there is no process by which this bread cast upon the waters will return to the same person who has set it afloat.

Our economic setup takes cognizance of large calculated risks, but it does not take cognizance of small uncalculated or incalculable risks. This is consistent with the fact that it is to the advantage of the community that the community as a whole, which means simply, some persons in it, should take these risks, but they are outside the normal specific profit and return setup. Who is to carry a new idea from conception to parturition?

In our community as it stands, we have scientific institutions, such as the universities and a few of the foundations, which are run on a non-profit basis. What is more, we have a few, but a very important few individuals in the community who are only moderately sensitive to the profit motive. When the important functions of these institutions and these individuals are taken over by industry, there results a very subtle but significant distortion of their purpose.

Certain industries and certain companies have, in fact, maintained highly individualistic scientists and highly individualistic sciences on their payrolls. To some extent, this is because the same individuals and sciences that contribute to the community at large and to the environment that a particular firm finds profitable may be kept on tap for more specific problems of development. In this double use of them, there is also the risk that they will find themselves swamped by these problems and cease to contribute to the science on a larger level.

Some individuals, and these sciences, may also make a certain secondary contribution to industry in the form of good name and advertising. Steinmetz and his association with the General Electric Company resulted in very important and immediately usable inventions, but it also resulted in conveying an impression to the public that something important to science was happening in the General Electric Company. This was not without its value to their advertising men and to their good name.

The case of Steinmetz was, however, a very special one. In addition to real top ability, his deformity and the individuality of character which was not altogether dissociated from this, made him picturesque. But a Steinmetz is a windfall, and those companies that have tried to make creative personalities of a less picturesque nature subservient to the advertising needs have had far from uniform success in the matter. An industry may shelter one or two men of this sort, but it cannot, in fact, shelter a sufficiently large and typical body of them to assure itself and the community as a whole of the continual replacement of ideas which is necessary to prevent the well of science from ultimately running dry.

There are other difficulties in the way of the commercial firm when it comes to the employment of the individualistic scientist, more especially when he is employed in the same building as more conventional engineers or development men. It is of the essence of the individualistic scientist that he must take his reward for being what he is rather in form than in money. If he is not that sort of a person to begin with, he will soon find the development laboratory and even the sales force a more congenial point of vantage from which to follow his main interests.

Yet this freedom which he has will also be a thorn in the side of the more conventional scientists who work in the same building, and even more a thorn in the side of the purely industrial employees. Unless he is a very eminent man, the unconventional scientist will be disliked. Even the eminent man only arrives at his position by going through stages in which he is not eminent.

Thus for general good will and good feeling, it is to the advantage of the community that the worker in the more fundamental research should have a home in which he is congenial and is wanted, and this home should often be an institution in which his love for freedom and the respect that he demands from his individual associates are not abnormal favors granted to him and to no one else, but a part of a natural environment in which he works.

Combine this limitation on his usefulness with the real doubt as to whether his services to humanity will accrue in particular to his employer rather than to some very different person who finds out what they are at a much later stage, and it will become clear why the commercial laboratory can never become the exclusive home of science.

The present age favors the big laboratory, and favors the man who is sensitive to the profit motive. Moreover, it favors the man who will meet his industrial colleagues at the country club or over the bridge table. The real scientist of the first rank is by the nature of his own activity too busy to care much either for money or the ordinary signs of prosperity with which we have been made so well acquainted by Thorstein Veblen.

It is true that the college professor must fight the peculiar conservatism of the college and the peculiar sort of social acceptance on which it insists, but although these restrictions on his freedom are real, they are certainly no larger than those restrictions which bind the industrial employee. The man who is too much of an individual for the college laboratory will almost always be much too much of an individual for the industrial laboratory.

There are other ways of killing a cat than smothering it in butter, but, after all, smothering it in butter is a very effective way of getting rid of it. There are other ways of destroying the intellectual fertility of the scientist than giving him a million dollars to work with and then expecting proportionate financial results, but that is also a good way of putting him out of action. To the able and conscientious man, the control of a large sum of money is a burden, unless it has a tremendously strong motive behind it. If I have nothing but my salary, I do not feel particularly humiliated if my new idea cracks up, and I go for a year or two without a success. If I were employed at the sort of salary which until recently was reserved for royalty and the heads of companies, I would very soon begin to question myself concerning my own sterility. It might well be that others would question me too, including my employer. At any rate, it would be difficult for me personally to accept a fabulous salary without forcing on myself the demand for fabulous work.

The free-lance does not fit easily into any particular niche of a purely capitalist system. To a certain extent, this has nothing to do with capitalism, for he would not fit any better into a large-scale Communist or socialist organization, which is compelled by industrial tradition or by state discipline to budget the unbudgetable, whether in money or in reserve national effort. The "megabuck" scientist, the man who plays at a million dollars a throw, is equally the slave of his position, whether the money comes from a private industry or a state monopoly.

I am making this book a plea for the effective scientist, who nevertheless wishes to lead the life of the small man and who, in fact, cannot fulfill his function unless he does lead such a life. I do not say that the capitalist system or Big Business or the Big State is universally invalid. I simply say that they do not represent a complete appraisal of the social needs of the community in the matter of invention and discovery. They all of them represent a certain localization of scientific activity and of scientific responsibility that is not necessarily for the good of the people, the community at large, nor even for the complete and long-time good of the Capitalism or Statism itself.

The profit motive may be important, but it must be supplemented by other motives. The community must cultivate a group that is neither subservient to the profit motive in the external community nor is internally governed by this motive. We need the free-lance in the arts and in the sciences, and the free-lance finds his natural home in institutions which are primarily designed to give him his proper scope. For the really vital stages of invention and discovery, the small laboratory and the well-organized foundation, such as the Guggenheim Foundation, which is tailored to the needs of the individual scientist, occupy an indispensable place.

The integrity of discovery and the possibility of a decent career for the discoverer and the inventor must be conserved by methods which are not confined to giving the inventor, his employer, his heirs or assigns, permanent property rights in thought. This is especially true if we consider, as many people do at present, that the rights of freedom of initiative and of personal ownership contain the implication that the proprietor or any chattel or right has an implicit license to waste, to destroy, or to suppress what is his.

Over history at large, as Professor Karl Deutsch has pointed out to me, there are relatively few epochs in which one's property and long-time assets are regarded as absolute, and ownership imparts the right to waste and to destroy the earned fruits of the land, as well as to conserve and to enjoy them. One of these periods was that of imperial Rome, another that of the colonial expansion beginning with the time of the Spanish conquests, and only now drawing to its close. Professor Deutsch has made it clear to me that these periods of unlimited property in land have also been periods of chattel slavery—that is, of unlimited property in human beings.

Most periods of slavery, and in particular that of the serfdom of the Middle Ages, have given the slave owner somewhat limited rights in his slaves. Serfdom attached the serf to the land, and assimilated him to the land as far as his ownership went. At least in theory, a lord had no more right to deal wastefully with the bodies of his serfs than with the soil of his estate. What further prevented him from owning his serfs or his land as his absolute property was that he held both of them with a certain sharp responsibility to the rights of his overlord.

The Roman *ergastulumand*, the slave camps of the Southern plantation, the Mexican silver mines of the sixteenth century, and the twentieth-century diamond mines of South Africa: all these belong to a tradition harsher and more wasteful than that of the feudal system.

Behind what I may call the plantation attitude, which permits one to work one's slaves to death and to grab the fruits of the fertility of the soil without maintaining the soil for future generations, lies a direct denial of the attitude of Albert Schweitzer: namely, respect for life. To treat human beings as consumable property assets is of course a lack of respect for life; but so it is to chop down a forest without replacing it, to squander the soil, or to render sterile the soil of the human intellect. It is because Schweitzer has seen the degradation which is common to all these, and has denounced it in no uncertain terms, that he has achieved the rank of a modern prophet.

Thus there can be no genuine ownership of a really fundamental idea but only a stewardship of such an idea in trust for the community. How can such a stewardship best be facilitated by measures which are fair both to the community and to the creative intellect? Without hoping to settle great matters of policy in one or two sentences, let me suggest one or two factors which may be significant.

In order for invention to fulfill its social function it must be recognized. It is not enough that somebody discover a new idea, nor is it even enough that this new idea be conserved in archives accessible to everyone. Information is not merely something which concerns what is said, but rather the proportion of what is said to what might be said.

The greatest book in the world may exist in a library. If there is no system of directing the attention of the reader to that book to the exclusion of other books, it is of little value; and of proportionally less value the larger the library is. In other words, the full merit of an intellectual discovery is diminished in the absence of a machinery for the recognition of this discovery.

The world must be full of mute, inglorious Miltons. They cannot perform the function of a Milton so long as they are mute and inglorious; or even so long as no one listens to their voices precisely because they are inglorious.

It is for this reason much more than for any titillation which recognition gives to the scholar, that it is important that inventive minds be selected from the ruck of less inventive minds by some act of recognition. No man is a real intellectual of the first rank who does not have a pretty good idea of the value or lack of value of his work. False modesty is not one of the major virtues.

On the other hand, a gem of purest ray serene is of very little value in the deep unfathomed caves of Ocean. It is not for the creative scientist himself that his recognition is primarily important: it is for the other scientists and for the people who are going to use his work. The principal valid point of a system of rewards and recognitions for the scientist is that it helps the direction of search of the learner and of the man who wishes to use scientific work.

It is neither desirable nor possible to give the discoverer of a brand new scientific idea the full monopoly of the commercial returns of his idea for a period of any length. On the other hand, there are ideas of such consequence that no man should have the returns which an effective monopoly of this kind would bring.

I am saying this, not merely for the sake of those who might be subjected to the scientist by this authority, but quite as much in the interest of the scientist himself. The maintenance of the position of being a Very Important Person, even if we assume the greatest internal modesty on the part of such a person, is a burden and an isolation. This is not least so in the case of a scholar, a man who should be up and doing something new.

Moreover, the value of a piece of scientific work only appears to the full with its further application by many minds and with its free communication to other minds. Here any secrecy or any rights or property possession will naturally have the effect of making people shy off from a preempted field of work. By this they will actually delay the intrinsic consequences of a discovery. As a secondary matter, they will delay the true and full recognition which the discoverer should have through the mass of work dependent on his own.

Since a purely property reward is not in order, there should be in my opinion some other system of rewards for scientific work of only remote and contingent practical application. This system should be directed toward the original (not the derivative) selection of the really creative individuals. It should aim at a certain facilitation of their work by allowing them to live in a satisfactory environment, which need in no wise be proportioned to the ultimate commercial value of their discoveries. Here I class discoveries and inventions together, because no exclusive property rights come into consideration. Discovery is as great a contribution to the future of knowledge and technique as is invention.

What forms such rewards should take, and by whom they should be administered, I am not prepared to say dogmatically. There are several alternative possibilities. Besides the prize, which represents an award limited in numbers and not adaptable to the changed level of production that is likely to accompany a universally increased scientific activity, I can think of the premium, in which the award depends more on the absolute merit of the intellectual work accomplished than on its relative merit in a limited competition. I can also think of the pension, which might either be awarded in such a manner as to give an immediate improvement in the scale of living of the beneficiary, or to cancel his fear of an indigent old age. I forebear to recommend one of these courses more than another, nor to pronounce upon the agencies which should be entrusted with this important social responsibility.

At any rate, whatever benefits are awarded for scientific creation should have the good of the community as their purpose, even more than the good of the individual. As such, they should be contingent on a full and free publication of the new ideas of the discoverer. The truth can make us free only when it is a freely obtainable truth.[2]

Notes

[1]Norbert Wiener (1894–1964) was the founder of the science of cybernetics, which is concerned with the common principles in the workings of computers and those of the human nervous system. In 1919, he joined the faculty at Massachusetts Institute of Technology, where he served until his retirement in 1960. A truly exceptional intellectual figure, Wiener was honored throughout his career for his brilliant achievements and often represented the United States at international scientific conferences. He left a long-lasting and increasingly growing intellectual legacy, including many current debates on the appropriate uses of modern technology, that is, on the relationship between technology and design, in his multifaceted treatment of the subject.

Immediately after World War II, Wiener grew progressively more concerned with possible dangers of technological growth going out of control, as well as with the social and moral consequences of automation. The first action he took, in 1947—after the shock he experienced in the wake of the Hiroshima and Nagasaki bombings and his pondering over how that decision had been made—was his refusal to provide any further research and information for military purposes. Behind his refusal was the assumption that in working for weapons design "the scientist ends by putting unlimited power into the hands of the people whom he is the least inclined to trust with its use. It is perfectly clear also that to disseminate information about a weapon in the present state of our civilization is to make it practically certain that that weapon will be used." That decision required him to his own embarrassment to renege on an agreement to participate in a military-sponsored Harvard symposium on high-speed computers. He found that he had placed a new barrier between himself and the majority of the scientific and technological community, an isolation that grew steadily. (Albert Einstein was an exception among scientists when he declared his admiration and support of Wiener's statement.) Wiener nevertheless tried until the end of his life to encourage pacific and civil-oriented uses of new technologies, without much immediate success.

He had asserted that his function as a self-respecting person and as an intellectual would be very different from that of an academically domesticated intellectual with primarily professional concerns. His style of behavior can be summarized by Lewis Coser's definition of intellectuals:

> Intellectuals are men who never seem satisfied with things as they are, with appeal to custom and usage. They question the truth of the moment in terms of higher and wider truth; they counter-appeal to factuality by invoking the "impractical ought." They consider themselves special custodians of abstract ideas like reason, justice, and truth, jealous guardians of moral standards that are too often ignored in the market place and the house of power.

The isolation of Wiener, then, was the result of his behavior, condemned as it was by his domesticated colleagues as much as by his own choice. He was and he remained throughout all his life the marginal individualist who could not and did not want to identify himself permanently with any institution or cause other than that of rigorous honesty with himself and with others.

Wiener did not share the faith in the future held by many Marxists; he had no systematic program for social and economic change, nor did he envisage a political revolution. For him, widespread adoption of wise and more humanistic "social patterns of design" was derived from an order of consciousness commensurate with the nature of the world, and he posed the issue more as a moral struggle, an affirmation of personal values.

The only special political force he recommended, and that primarily by personal example, was conscientious noncooperation short of violation of the law. Wiener espoused this philosophy despite its result in personal loss of income and exclusion from scientific circles. Wiener's overall social theory formed a coherent whole that placed value primarily on human life and development and not on technological "progress" or political power and control upon it.

[2]Wiener's writings include *Cybernetics: Or Control and Communication in the Animal and the Machine* (Cambridge, Mass.: MIT Press, 1961); *The Human Use of Human Beings, Cybernetics and Society*, a two-volume autobiography consisting of *Exprodigy: My Childhood and Youth* and *I Am a Mathematician: The Later Life of a Prodigy* (both published in Cambridge, Mass.: MIT Press, 1953 and 1956, respectively); a philosophical novel, *The Tempter* (New York: Random House, 1959); and the essay *God and Golem, Inc.: A Comment on Certain Points Where Cybernetics Impinges upon Religion* (Cambridge, Mass.: MIT Press, 1964). This last book was posthumously awarded the National Book Award in the category of science, philosophy, and religion. The four volumes of his *Collected Works* have been edited, with commentaries, by P. Masani (Cambridge, Mass.: MIT Press, 1976–1985).

See Heims, *John von Neumann and Norbert Wiener: From Mathematics to the Technologies of Life and Death* (Cambridge, Mass.: MIT Press, 1980). A discussion of Wiener's role in the design-technology-society debate can be found in David Noble, *America by Design* (New York: Knopf, 1977) and *Forces of Production. A Social History of Industrial Automation* (New York: Knopf, 1984). For detailed biographical information, see Nathan Levinson, "Wiener's Life," in *Bulletin of the American Mathematical Society*, 72, no. 1, part III (1966). This special issue is dedicated to Wiener "in recognition of his towering stature in American and World Mathematics, his remarkably many-sided genius, and contributions to Science." Marco Diani is also writing an intellectual biography of Norbert Wiener, tentatively titled "The Double Bind of Innovation: A Life of Norbert Wiener."

Communication Esthetics, Interactive Participation, and Artistic Systems of Communication and Expression

Fred Forest

Translated from French by David Sugarman and Joanna Weston

Communication esthetics directly envisages transposing the perceptible principles that are observable in the evolution of the environment and of our world onto the function of art itself. Therefore, this function should no longer be considered in terms of isolated objects but in terms of relationships and integration: works of art, information, and art systems must all be perceived as integrated wholes—wholes that cannot be divided or reduced in any way to the sum of their separate material parts. What constitutes the "work" is no longer its material medium or its visual or pictorial representation but precisely that which is not perceptible by our senses but only by our awareness. In generalizing the methods of production of images and in making the process commonplace, our society has limited its esthetic treatment of them and has transferred legitimate artistic intervention from the production process to the invention of models. The inflation of images has inevitably led to their devaluation. Esthetics now seeks its favored ground elsewhere than in the incarnation of the plastic sign. No longer able to operate on the method of representation, the artist now intervenes directly on reality; that is to say, he carries out his symbolic and esthetic activity using different means from those he has used up to now.

The approach in which I am currently engaged is work that has communication itself as a goal. It consists of not only thinking about communication but also of practical activity in and around the field. Such a position throws all of the traditional data on artistic activity into disorder and makes the perception of them problematical. We are witnessing not just a change in the object of art but also in the means of achieving it. All of this leads to a profound modification of esthetic philosophy. Through a range of experiences, *sociological art* supported the existence of an art of action: an art of action whose programmed development was situated in a social space and took into account the environment into which it was born. Based on a theory of actions, it acted on the world in order to bring about change. It brought communication theory into play by producing a process of interactions between individuals or groups of individuals.

This type of art functions as a transmitter of original messages, which are both specific and disturbing. Artists take up the position of the sender of the messages; they speed up and activate communication; they innovate, either by introducing parasitic messages into established circuits or by setting up their own parallel networks. Sometimes this is achieved by setting up intersections and connections between them. Such action immediately results in a critical use of pervading information and overloads the routine function of such specialized circuits. It must be emphasized that the novelty here is found in the transfer of the field of action of artistic practice. Communication artists generate symbols just as the traditional artists colonize other realms and annex other fields of endeavor. They are not content with preestablished places and circuits reserved for this particular use and for a particular public but deliberately transfer their production to other fields and channels. By transmitting through mass media rather than through art museums, the messages have less specific targets, but the target the art museum aims at is nonetheless hit through this channel. In any event, this can only widen the circle of potential recipients, reach them from afar, and in this way achieve a new type of relationship with them, encouraged by the originality of the situation thus created.

Communication artists introduce their own signs, which not only work through the daily communication media (newspapers, radio stations, television, and the telephone system) but are also "about" them. They are juxtaposed with societal signs, also vehicled by these same channels. Thus communication artists are operating on the space of their time, which is the space of information. Into this space of information artists are insinuated and install and stage their symbols. Of course, as a result of the chosen framework of action, the communication strategy employed will dictate the choice of medium, timing, and type of organization in function of the message to be put over and the goal to be achieved. By appropriating other channels in this way, artists also point out the thoroughly relative space that up until now has been allotted to artistic creation in our society, isolated as it is in highly localized preserves. Today the field of information is opening up an unlimited space for action to artists who are capable of inventing specific forms of art.

The practice of sociological art has always drawn particular attention to communication problems. Certain detractors have accused it of inflating information through its own working, particularly with reference to the activities of the sociological art collective. This is something specific to the methods of sociological art following its logical thread. This communication has been expressed in various forms. Artists have had recourse to media appropriate to the moment and to the circumstances.

For understandable financial reasons, the mailing of documents was the most widely used of these means. Obviously it allowed for the saturation of the public to whom they were addressed; above all, "arbiters of taste" could be reached, and they in turn relayed the information. Nonetheless, our occasional actions through the mass-communication channels of press, radio, and television were numerous and well remarked upon. Both the dynamic and the amplification of information were part of the dimension we consistently gave to our work. Major information channels allowed us to endow our events with the immediacy and the social impact we expected of them. We always gave careful thought first to the preparation of this information and then to its circulation. In all of our actions, the technique of communication was the subject of prior in-depth research, the object of an exhaustive plan. Although this was an integral part of our work methods, the plan of action had to be flexible enough to allow for adaptation to any unexpected situation. Our action on the *artistic square metre* was exemplary in this respect, as shown by the results achieved at the time. Without a doubt it is here that an ability to intuit both the media and an awareness of their operating procedures comes into play. This awareness is borne on the air of the times, induced by the informational environment into which we are all plunged.

Games are activities performed freely, with no obligations and for pleasure. As such, they form one of the most fundamental components, in the widest sense of the word, of all artistic events. This certainly does not mean that art is a gratuitous occupation with no determined objectives. It is not merely escapist entertainment that tends toward fiction. Art maintains close links with reality and seeks to use its influence to modify the perception of it. From the field of possible situations, games as simulation models anticipate real situations through successive investigations. Games develop strategies for action; that is, they help to redefine social relationships and behavior by reproducing them on the level of game playing. They suggest alternative relationships and behavior; they modify them. In this guise, art operates directly on social reality. It posits a simulated representation of this reality, and the imperfections of the reality are highlighted through the juxtaposition with the simulation. Culture is no longer satisfied with being only a leisure activity: It now wants to assert itself as a fighting weapon.

According to Marshall McLuhan, "Any game, like any medium of information, is an extension of the individual or the group. Its effect on the group or individual is a reconfiguring of the parts of the group or individual that are *not* so extended. A work of art has no existence or function apart from its *effects* on human observers. And art, like games or popular arts, and like media of communication, has the power to impose its own assumptions by setting the human community into new relationships and postures. Art, like games, is a translator of experience. What we have already felt or seen in one situation we are suddenly given in a new kind of material."[1]

The conception, organization, execution, and very goal of our artistic actions all attempt, by using the appropriate methodology, to bring fictitious situations and real facts into contact with each other. It is as a potential "other reality" that the fiction is presented to the real world, a reality that is enriched through the shared experience of contact between artist and spectator. Games, dreams, and the imaginary are brought into the dimension of lived experience. Such a conception of art clashes completely with traditional codes and makes its perception problematic. In the field of the plastic arts, the works of centuries past generally followed the rules in order to produce a certain "verisimilitude," and it was this verisimilitude that was the main criterion on which judgment of them was based.

Every truly innovative act necessarily involves a break from established order. Fundamental artistic innovations must always draw on the repertoire of established knowledge but are nonetheless enriched by each artist's creativity. The brutal eruption of new idioms into the cozy world of art inevitably entails the natural phenomenon of bewilderment on the part of the general public and so demands a period of assimilation. In the current broadening of the artistic spectrum to encompass disciplines belonging to the social sciences, personal expression is likely to become the reflection of a more general problem and all its implications, be they political, social, psychological, or philosophical. The integration of the social sciences into the context of the plastic arts is accompanied by a diversification on the level of techniques and borrowings from literary genres (narrative painting), as from the theater (happenings), the cinema (video art), and other fields. McLuhan writes: "As our proliferating technologies have created a whole series of new environments, men have become aware of the arts as 'anti-environments' or 'counter-environments' that provide us with the means of perceiving the environment itself. Art as anti-environment becomes more than ever a means of training perception and judgment."[2]

For a very long time discourse on art consisted essentially of discussions about numbers of angels on the head of a pin. Things are starting to change. Through his work the artist is today starting to understand that "power" is linked to every human action. Attempting to deny this, in the name of some naive idealism, is tantamount to denying reality. People are surrounded by constraints and enjoy certain liberties. The relationship between people is always conditioned by the power game constantly played out between them. There is no reason to flinch at recognizing it. Power can be seen to be operating at all levels of human relations. It is the attribute of every social performer. Everyone exercises power, at the same time as submitting to it. Each one of us has been forced to "reconcile" himself with his environment since earliest childhood. Each one of us has found it necessary to elaborate a behavioral strategy, whether consciously or unconsciously, inside the system in which we operate.

Individual and collective change require overthrowing the rules of this particular game. Each one of us has to learn to challenge the constraints and liberties that constitute our "field of action." It is precisely because it took account of these facts that sociological art thought of itself as an "art of action." Even in the most rigidly controlled social systems, there is always a margin of maneuverability into which either an individual or a minority can manage to slip. Therein lies hope. In a trial of strength, the weakest are never totally defenseless but always have the means of turning the situation to their advantage if they can find the exact spot at which to apply the lever. The ideas of "game" and "strategy" are closely linked to social behavior. Its limits are, of course, those of opposing authority, but also those of our own imagination, which requires exercising, stretching, and sharpening. In turn, artists become "social operatives"; they become social performers. The scaling down and the provocation of power, and its recuperation as a form of play, belong to the field of art. Responsible artists know this power as theirs and confront the surrounding world with it.

Modern communications had their beginnings in the latter part of the nineteenth century with the development of the electrical manufacturing industry; computer technology has now brought us firmly into a communication society. These technologies are at the heart of changes in social reorganization during the past century, and they have modified not only our physical environment but our mental system of representation. Electricity, electronics, and computers also provide artists today with new instruments of creation. All of this is why we must constantly reconsider our perceptions in order to apprehend the world in which we live. On this level the artist has something to say, something to do. Throughout the ages, the successive emergence of new technologies (for example, the technologies of raw material transformation, energy harnessing, and, most recently, information) has involved people in varied and successive forms of expression. Contemporary awareness is molded through the multiple channels of the mass media. The previously prevailing notion of "art for art's sake" has been called into question. Today's artist, and more precisely the communication artist, reintroduces esthetics in its original anthropological function as a system of symbols and actions. A new esthetics is in the process of emerging: Communication Esthetics.

The very word "artist" necessitates some adjustments in a society that is undergoing mutation. The roles, the means, and the awareness the term denotes are evolving. It is imperative for "artist" to become disassociated from the ideological connotations that still link it in our minds with a romantic and anachronistic vision of art. There still exists a gap on the political and educational level between "acquired culture" and "culture in creation," and it has perhaps never been so noticeable as it is at the present time—in the computer and television age. Stricken with vertigo and anguish before a changing world they are unable to come to grips with, humans have a tendency to seek refuge in the past.

Artists refuse this retrograde vision. They face up to the present, pushing themselves to explore its possibilities. Artists also observe and are involved in the adventure of their own epoch. They can neither ignore nor escape from the radical changes currently shaking its foundation. In their roles as artists, they are faced with the imperative task of grasping its "meaning" and formulating its "languages." The intention is not, of course, to challenge the scientist and the technician on their own ground. This would be stupid and naive. On an altogether more modest level, the intention is to use, even to "divert" the new tools of knowledge and of action in an attempt to widen the horizons of our perception, of our awareness, and of our consciousness, in order to revive our codes, our ways of seeing, of thinking, of understanding. In the same way, to allow the individual to find his place, here and now, in the world is no simple undertaking.

More and more, the concept of "relationship" plays a key role in contemporary thought. The idea of relationship is present within each science and is central to an ongoing interrogation about science as a whole. Any individual is part of a complex network of interrelationships

FIGURE 1.

such that everything has something to do with everything else. This notion has an important place in many branches of science and pervades our awareness. Art refuses to be excluded from this idea of systemic arrangement. The fact of communication relationships is the hallmark of our time. Such fields of research as cybernetics, information theory, games theory, and decision theory all have natural links with the preoccupations of artists who are particularly attentive and receptive to the "wavelengths" of their age.

It seems to me necessary today to knock down disciplinary compartments and reinsert art into the systems situated at various levels of the organization of reality. In our society, artists inhabit a multiplicity of specific times and spaces. Their life and work are part of that complex network we spoke of where everything circulates in all directions along different connecting circuits. Today, before they ever deal with "content," artists must express these connections, along with speed, nature rhythm, flux, and the data that flow both through them and through us. Although not always recognized as a prime investment in our utilitarian society, art, too, has its rights and makes its demands, just like the sciences, technology, and politics.

It seems appropriate here to develop at length some remarks on the nature of the relationships that link art to its entry into computerized

FIGURE 2.

society. This is not with the intention of considering any specific problem, such as the effect of computer-generated images on creativity, manufacturing production, and the resulting economic structure, but to remain at a more general level, a somewhat more philosophical one. The relational aspect, of which we are not always conscious and which is about to affect the art world directly, is of prime importance.

Having lived through various production societies, we are now in the communication society. Even if electricity, electronics, and computer technology have provided artists with new creative instruments today, one cannot help but notice an enormous resistance within the social body to all change. This resistance is particularly felt in specialized art circles and institutions where the prevailing mentality is frequently that of an earlier century. Outside of the market place, a few artists nonetheless doggedly pursue fundamental research, despite a nostalgic fashion in art that is constantly advocating an unquestioning return to painting. By giving preeminence to pictorial pigment, the current art market is only responding to short-term economic imperatives. Tangible objects are, of course, essential to supply the commercial art market. Unless information has first been made material and tangible, dealers have not yet found a way of making it part of their capitalizable merchandise.

Telephonic stock exchange information has become an economic "object" in itself (for a stockbroker), just as have erotic telephone calls charged on a fifteen-minute basis (for others). It seems that artists, not to mention painters, have a long way to go before their productions can be sold in this way. This, of course, is due to the fact that art, contrary to applied science and economics, has no practical application whatsoever in everyday life. It must be behind the times. For the most part, it is, unfortunately, considered purely "ornamental." The "pressure" from our surroundings is not, however, without an effect on the very nature and type of artistic production. Despite the extremely slow rate of change in the art distribution and consumer circuits, notable evolution has taken place. Different stages have taken us from the esthetics of image to the esthetics of object, and from there to the esthetics of gesture and of the event (the happening). This trajectory shows a slow "dematerialization" and "disintegration" of the art object.

The concept of communication was already the central idea of sociological art, the first activities of which I carried out as early as 1967 and the principles of which I first expounded in 1969. I have always considered that the natural field of artistic action is the terrain of social activity, a field that may be enlarged and explored thanks to the new communication technologies. This option upsets those with a fixed concept of esthetics, who are incapable of grasping the obvious articulation among this new practice, the concept of art, and a society in transformation.

We are called upon to ask the question: Where are the frontiers of art situated? It is a brave person who will stick his neck out. There is no frontier. Art is an attitude—a way of relating to something—rather than a thing in itself. There is an esthetics of behavior, an esthetics of gesture, just as there is an esthetics of object. We have now to take a new category into account: the esthetics of communication. The media of this esthetic is often immaterial: Its substance comes from the impalpable stuff of information technology. In the sky above our heads, the electric signals of this information trace invisible, blazing, and magical configurations.

In the systems of reciprocity and exchange that are set up by communication artists, public participation cannot be overlooked. In my view, it will come to take on an ever-increasing importance. In the 1970s it was supposed that public participation would only take the form of a collective, and necessarily physical, relationship. These types of action, while well intentioned, soon fell into the context of "community art action," out of which some artists have never quite managed to pull themselves. What I have in mind are more involved forms of participation, such as those that occur through multimedia exchanges of information set up by the artist, who is present as the conceiver of the

FIGURE 3.

system and possibly also as the actor-animator of the whole. This idea of feedback and reciprocity as advanced by cybernetics has already, outside the world of science, found an application in the most ordinary of our everyday activities. These are the kinds of practices that sustain today's awareness and contribute to its formation. It is this contemporary awareness that to my mind is absent from the theater of operation of the plastic arts.

Traditional form is over and done with. There is a marked tendency for a more global culture in which the distinction between the categories of science and artistic creativity loses its meaning. A new definition of the triangular relationships among artists, theoreticians, and spectators necessarily gives rise to new esthetic thinking. A new art is being born, based on the aspirations and the creative needs of humans, and, consequently, encompassing their environment; it is an art able to go past the level of conceptual art, as well as that of propaganda art. Despite the diverse nature of its origins and of the forms it takes, the art of environment has a unified direction. By implication it tends toward a wider dimension, that of an authentic "sociological space" (a privileged area of investigation).

Just a few years ago, this concept of space was linked to the idea of physical representation, geographically defined. The proliferation of all kinds of media and their widespread utilization has led me today to a more "abstract" concept of this space. This is the "encounter" space built upon communication media. It is the space of social communication, created by a superimposition of all the technological media on our physical space. The idea is that of an immense chain-mail network, made of invisible wires, which convey our messages and through which our emotions are exchanged. Into this netting are woven new kinds of relationships between human beings, thus opening up an extra "reality"—a "mediatization" space that must be seen more and more as a new and privileged field for contact and interactivity; a dialogue territory dredged, like so much reclaimed land from the sea, out of the void by communication technology. Environment itself has a tendency to "dissolve" and resurface as an area in which our relationships become "tangible" by means of information. This more abstract sort of environment is no less real either in our representations or in our experience. Formerly the mere mention of the word "environment" made us think exclusively of perception of our physical surrounding. This has been particularly the case with architecture. Today this has evolved, and the concept of space is more and more associated in our representations with the idea of "information environment."

Artists have plenty of soil to turn in what is still virgin territory for them. They have yet to contribute through their practice, their thought, and their imagination to the creation of the fundamental precepts of an art built on communication: a communication art that irrigates the networks with a flood of data of the imaginary. The communication artists employ telephone, video, telex, computer, photocopier, radio, television, and so on. They do not simply use them one at a time, but they arrange them into systems and installations. From now on, this is how their capacity to create and invent will be brought into play. They make up given configurations, networks of varying complexity, within which they position transmitting and receiving multimedia equipment. They organize it into interactive systems, which they then animate.

The communication artist has become a sort of information architect, who sets off processes in an interactive relationship of participation between interchangeable partners. "Diagrams" or "information architecture" assemble and dissolve. They can, at any given moment, become the subject of a "photograph" that freezes them. The fulcrums of the artist's network are not fixed points that are just technical or formal: They are anchored and directly connected to the social fabric. Information technology facilitates interference between compartmentalized sectors. For the first time, artists can now hope to express themselves in fields other than those restrictive ones previously assigned to them. It is highly probable that the key idea of "bringing into contact," which marks our thinking and practice today, will become a preoccupation of artists and will figure ever more significantly in their creation in the years to come.

The overproliferation of visual media and the booming growth in the number of images they produce contribute paradoxically, if not to the disappearance of the image and its esthetics, at least to its devaluation. This suggests an explanation for the shift toward a new form of perceptual behavior latent in society. It is this the communication artists seek to integrate into the realm of art and this they seek to organize into the new framework they are advancing: Communication Esthetics. Writing about what he calls the "oversaturation of the world" by the image, Jean-Luc Daval points out: "Those whose function it was to produce the richest and the most meaningful images had no alternative but to disappear or shift their field of practice. It is this that explains why the creators of today need to produce new images far less than they need to know what to do with them, drawing upon their power of communication and of contact. At this stage of cultural development, the work of art must change its function. Henceforth, instead of conveying concepts or ideologies that are exterior to it, it must rather call into question its own status, the elements which constitute it, and its power of relation. When the media freed the image from the exemplars found in museums, there was nothing left but for artists to put it on trial, and relativize it completely. The question of the relational in art is going to have to be put differently from now on."[3]

In Umberto Eco's notion of the work as a structure,[4] we see the ideas of system, the arbitrary, and the implication of the spectator in the process of communication as advanced by artists. In the new self-assigned role of communication artists, they no longer present themselves as the "manufacturer" of a material object but base their approach on the particular, specific, and original relationship they establish among themselves, the spectator(s), and the environment. For the sake of clarity, it must be repeated that this kind of approach can in no way be assimilated to the kind of creation that stems from conceptual art. It is true that the communication artist, like the conceptual artist, relies on one singular idea, but its presentation owes nothing to what might be termed abstract "beauty" in a formalized setting, destined solely for the well-targeted art museum or gallery. Works that stem from the sphere of communication and that invoke its esthetics give rise to the concrete and operational installation of a materialized, functioning system, even though it may be that, spread out in space, the whole system cannot be entirely taken in at first view.

Observers can always notice the presence of certain elements (physical) or signs (visual or audible) that, by a process of mental projection, lead them to reconstitute the overall representation. They can intuit the representation of the positioning and of the relative placing of its various elements in a space that itself has different levels of reality (geographical, extensive, social, or communicational space); it is also the representation of the flood of information and of its configuration in the movements that bring it to life. By proposing systems of communication as "works," discernible in terms of their functions and their motions, the communication artist claims quite simply to be modifying our habits of perception, claims to have an effect on our perceptual behavior and on the very interpretation of art.

Now that they have become the conceivers of information exchange systems that they design and animate in a social communication space, artists' status has changed. In the past they "manufactured" objects rather like craftspersons, sometimes more like an industry. Now, though, art has lost its materialism once and for all: The artist "produces" a service. This evolution corresponds perfectly to the curve of evolution in society: It has been transformed over several decades from a society of production to a society of exchange. Art as practiced by the communication artist is the art of organization, which is no longer concerned with objects but rather with functions.

We noted earlier the emergence of successive technologies throughout the history of humanity: the technology of raw material transformation, the technology of energy harnessing, and today the technology of information. Unquestionably, these stages have conditioned the development of certain forms of art at given moments and will continue to do so. The most recent stage, the technology of information, no longer produces objects but pieces of information that are organized into messages or into "communicational situations." Art now transmits, receives, organizes, and diverts information and messages. Hence it must lay the foundations for the new Communication Esthetics and can be regarded as a reflection on the nature, the circulation, and the representation of messages in the social communication of our time.

To be confronted with a "work of art" in the organization of meaning produced by dealers, museums, and collectors is thus, first and foremost, to be confronted by the system of exchange that upholds it. "Organization of meaning" must be understood as all the reflexive arrangements in which the existence of each element is justified and legitimized. It is for this reason that once the constituent and dissolving functions carried out by the media and by the art system in its relationship to the "artistic message" have been recognized, the artist's interest becomes completely detached from the messages themselves, in order to focus on the techniques and social mechanisms that produce them. This means that instead of continuing to dwell on "information" and "meaning" as artistic research has done, or has thought to do up until the present, the artist is now in a position to be able to thematize, invest, and represent both information-free communication and systems of meaning without "signification." The problem being examined here concerns not only artistic production but the entire universe of communi-

FIGURE 4.

cation as well as the whole system of exchanges. Everything, in fact, can be subjected to investigation and consideration of an esthetic nature: The relevant area for esthetic research from now on must be considerably enlarged and extended to technological as well as social media.

Even though the idea clashes with our humanist heritage, the new technology is progressively modifying our value systems, our thought systems, our perceptions, and our sense of time and space. It is not at any time the aim of Communication Esthetics to deliver a naive apologia exalting technological prowess. Unlike certain artistic movements ("Futurism," for example), Communication Esthetics draws attention to the danger of the use of technology growing in a way that is completely divorced from any ethical, philosophical, or social consideration. Communication Esthetics has come forward with the ambition of working toward a new apprehension of reality and supporting a conception of the world that takes us further toward deeply spiritual goals. At the very moment when Eastern thought in all its forms is exerting a fascination over an ever-growing number of people, an active scientific elite is revealing that mystical thought in fact provides an adequate framework for contemporary scientific theory. The imaginary sense and strained questions as to the meaning of existence remain unchanged since the origin of humans. The most burning questions of the moment are still the mysteries of life, death, love, anguish, and pleasure. It is rather the way in which these questions are asked that is changing. Contemporary artists see themselves as equipped with new methods of investigation for exploring the collective unconscious and giving it form. Technological resources take them into unknown territory, which it is up to them to explore. The real stakes for contemporary art are now placed well beyond the status of the image and the status of form. We are now playing with the relationship we embody in our contact with the world—otherwise known as reality. Against a background of esthetic behavior patterns that are changing along with technological evolution, artists taking up these new instruments are suggesting the constitution of new anthropological models.

Time and space will constitute the artists of tomorrow's "raw material." Just as down through the ages they have worked stone, marble, wood, or metal, they must now attempt to leave their mark on these "immaterials." Time and space are not just physical concepts, currently seen to be evolving considerably with the progress of knowledge, but also realities capable of being lived. It is on this terrain that artistic practice can take place and be legitimized. In the unconscious minds of Western humans, the notions of time and space are indissolubly linked. We, as Westerners, do not have the slightest doubt that time and space are organically structured. Three-dimensional space imposes itself as an imminent fact in the world.

As for time, its linear unfolding accompanies us everywhere: With the past behind and the future ahead, we advance in the present. Human-kind builds its temporal horizon along a line of progression, the three separate zones of which are delineated by solid but movable cursors. Until now, this linear awareness of time has appeared as basically inbuilt. The new concepts that science is putting before us, just like the daily use of new technology, are calling these mental schemas into question. Our "certainties," acquired from and based on past sociocultural data, may well need to be rapidly updated. The new structure of time has already produced some spectacular social effects. In medieval times church bells tolled the hours; Frederick Taylor's time-and-motion stopwatch made production work to within a second; today the microprocessor allows us to control processes measured in nanoseconds. Microelectronics is defined as a new structure of time, the fine gradations of which go beyond human perception. What this really means is that if yesterday we could hear the ticking of a watch movement or actually see the swing of a clock pendulum, today we require a vast leap into the imaginary to understand how a calculator works.

By structuring physical space, the automobile has given us mastery over distance. Its appearance on the scene totally transformed our natural surroundings, our economy, our way of life. Transformations of an even greater order are in view with the arrival of the computer. It has managed to colonize us and restructure, in irreversible fashion, our time and space. The computer will soon be capable of bringing about the synthesis of technological and symbolic thought. The steam engine was a beneficial substitute for the resources of humans or animals: Computers amplify humans' intellectual resources. The current development of computers demonstrates that, in the long run, it is the machine that will allow us to return to our greatest myths, return to them to the extent that it will contribute to pushing back frontiers that time and space have always imposed on humankind. This particular development in computers is expressed by a sharp increase in speed, that is to say, by their increased capacity for real-time operation. The hooking up of computers to each other, and also to other machines, is a forerunner of the opening out of the telecommunications network and the abolition of certain constraints of distance. The distribution and multiplication of centers of decision, in leading to the dissemination of knowledge and power, give us hope for new forms of sociopolitical structures. Actually, in the light of this, we are witnessing a new recognition of the emergence of individuality.

The so-called "fifth-generation" computers, lurking on the horizon a few years away, are going to bring us into the as-yet-unknown universe of artificial intelligence. Not only will they process data, figures, and letters but also knowledge, through the development of deductive reasoning. The difficulty of mastering a new means of expression, whether it is canvas and paint or the resistance of marble, has always played a role in enriching the creative art. This essential enrichment will come less, perhaps, from the facilities that computer resources offer the artist than from the difficulties they present when he uses them to express his awareness, difficulties that will involve him in uncharted ways.

Are we at the dawn of a new cultural renaissance? Will telecommunication technology create the objective conditions for an alternative form of being together, on a scale that does away with physical distance? In all domains, the act of creation is freeing itself from spatial and temporal constraints, thanks to long-distance transmission, to the gathering of data through message networks, to the possibility of meeting without physical travel, and so forth. We must also take into consideration that the amazing calculating capacities this equipment possesses can allow artists the hitherto unequaled power of astonishingly rapid exploration of the infinite field of possibilities raised by the world of dream, of the imaginary, and of human thought.

This is the aspect of the transformation of our relationship with time and space given prominence by artists who claim their place in the "community of awareness" constituted by Communication Esthetics. In their varied work, they all share a concern for questions referring to spatiotemporal dimensions and to chronotopological realities, questions that have never been as acute as they are today. Possibilities for geographical travel are opening up more and more rapidly, our capital of information is widening, scientific experiments are being performed on time, and all of these are causing us to vacillate more than a little in our strongly held convictions on this subject and on a good many others. These falterings provide artists with a historic opportunity to bring about a rupture in the conventions of representation, thus setting them on the path of the extratemporal phenomena that truly constitute the fundamental problem of our age. Moreover, signs can be seen of a convergence of modern awareness and the profound and ancient sources of religious, philosophical, and mystical oriental thought. We can but remark that all these transformations brought about by media systems are, without our knowing it, reorganizing our whole system of esthetic representation.

Let us emphasize the fact that contemporary awareness is shot through with doubt and uncertainty. Well-established notions of space, notions of time, scales of size, are now all called into question. Our age is going through a profound crisis of perception at a time when the theoretical interpretation of various physical phenomena is being challenged. Our awareness is therefore marked by the surrounding climate and molded by the continuous effects of the fundamental changes occurring at an ever-accelerating rate. For modern humans, the technological media have become artificial extensions leading them into the realms of time and space, which were still inaccessible to them only yesterday. Television and the telephone daily send us to the other side of the globe and bring the world into our living rooms. Instead of the traditional concepts of the distinct object, limit, and unity of place, we must now rather react to concepts of network, commutation, and simultaneity. Ubiquity is no longer a utopian vision of the spirit; under certain conditions, communication technology produces it every single day.

The age of awareness in which we live is recognizable by the transfer of information and by the dynamic configurations that catch us up in their movement, representations that come to life in structures with interchangeable elements known as installations, systems, networks, and the like. The resulting strongly felt changes in our perception, in our relationship with the world, and in our everyday behavior attest to the birth of a new esthetics. It is an esthetics whose designated object lies beyond the visible, beyond the tangible, in the zones of infraperception, alongside our modern awareness.

The technological systems of exchange in which we are directly implicated, both as active participants and (either individually or collectively) as constituent parts of the system, open the way to awareness

FIGURE 5.

relationships that no longer must pass through visual and verbal channels. Our daily living experience unfolds in a global field of interactions and events created by electric and electronic media. The permanent information bath in which we live reorientates our ways of feeling into new forms. Inevitably, if art had not been deflected and diverted from its natural vocation by economic pressures, it would have been able to fulfill the expectations of the new awareness. Art could thereby endow awareness with its own specific forms of expression, discovering them as it went along—forms of expression that precisely spring from Communication Esthetics and not from a type of art that is something to see or listen to, a type of art whose practice and whose objectives lie beyond the image, beyond the pictorial act, beyond the object: in communication itself and in its modes of functioning.

The process of dematerialization of art that goes back to Duchamp and the recourse of artists to concepts, attitude, and intention encourage an open reading of art. It is the very domain of art that is widened. Yves Klein's school of awareness, his theater of Void, and his cosmos perspective prophetically introduced us to a civilization confronting the conquest of space and the mysteries of matter. In the age of electronics and telecommunications, humans are making their way further and further toward a less concrete relationship with reality, toward the dematerialization of their everyday experience.

Our awareness cannot but be profoundly changed. All that art can now make us "feel" is that patterns of sound and encoded images are only illusions, behind which millions of electrons are agitating. What Communication Esthetics artists endeavor to "represent" is a representation that draws its origins from beyond the real, beyond appearances, and beyond the normal perceptual framework. Technology involves us in "data acquisition" of the world in which the physical marker has lost its meaning in favor of electronic sources of assessment. Representations on video and computer screens substitute so powerfully for the materiality of distance that they are on the verge of doing away with their referents.

The foundations on which yesterday we claimed to build and legitimize our representations have become precarious and often suspect. In the case of the television image, for example, our perception vacillates under the temporal shock of the instantaneous broadcast. In this image, the physical obstacle, like the obstacle of time, suddenly dissolves in a blue cloud of electrons. Space is flattened out, truncated, eroded by the vector of communication. The fast forward, the slow motion, even the rewind of the film or video image overthrow our concepts and our conventions about time. The Euclidian heritage of the notion of space as continuous and homogeneous crumbles away before the new concepts of discontinuous space, a space dotted about with markers that our perception on the human scale is quite incapable of picking out. From now on we must, therefore, learn to inhabit the temporary. We have to get used to the idea of permanent wandering, adapting ourselves to an instability that we will eventually have to tame. In the end we must find the point, at once fixed and moving, from which our vision can discover and invent this new relationship among our lived-in space, our electronic space, and our space in evolution. In order to do this, we must rely on and integrate as rapidly as possible notions bearing strange barbaric words for names: commutation, arborescence, intermittence, modular, interactive.

The ideas of, and the work undertaken by, Communication Esthetics help us to share and understand processes that are still complex. Through the artists who represent it, Communication Esthetics helps to bring to light the sensory contact we have with the new media. After having thought for a long time that these new media have "desensorialized" communication, we have now to admit that they do nothing of the kind. They have become integrated into our lives to a greater and greater extent; they now constitute a sort of sensory network through which our exchanges are constantly traveling. They have become the bases, the extension, and the amplification of our most intimate vibrations. Our dependent relationship with communication technology in everyday life allows us to confirm that this situation is giving rise to a new form of awareness. Television, for example, has created a singular form of esthetic relationship based on the "distant presence." Television, like the computer, is a source of strong environmental pulsation; we have not yet mastered the effects of this on our nervous systems. Questions can indeed be asked about the way in which long-term use may even transform certain of our thought processes. It is apparent that the media systems in our electronic society heat up our "cold" environment and bring with them a certain degree of "sensualization." We are permanently immersed in an electronic bath that dispenses an increasingly intense range of stimuli to the individual. The body of society, like our own body, is caught up in a huge net of communication.

Now I wish to address myself to those who point out the risk of our being cut off from a direct physical relationship with the immediate world, those who have genuine fears and a keen sense of nostalgia. As of now, hybridizations, which constitute rites of passage, are being carried out. More and more these hybridizations are bringing humans into closer association with the machine. In the not-too-distant future it is highly foreseeable that the computer will play the role of interface between technical and organic functions. Electronic media are bringing about a cognitive rupture that constitutes a veritable psychological revolution and that may well radically modify our relationship to the world. Contrary to the most pessimistic fears, this revolution is enriching the sensorial faculties of our organism. Our tactile and acoustic senses are being actively solicited. Perceptible facts and cognitive facts are from now on simultaneously integrated into new configurations that cannot be contained by linear thought.

The communication artists attempt to show through their actions that we are situated in the center of a global information process and that its complex mode of functioning places the individual in a brand new position from which he is obliged to discover and invent new forms of adjustment to his surroundings. The goal of communication artists is certainly not to produce first-level meanings but, above all, to make us aware in the end of how the generalized practice of communication interreacts on the whole of our sensory system. This evolution is about to put into place the data for a new awareness at the edge of our perception, and then, along with new ways of feeling, it will open up new esthetic paths.

Notes

[1]Marshall McLuhan, *Understanding Media* (New York: McGraw-Hill, 1964), p. 242.

[2]McLuhan, *Understanding Media*, p. viii.

[3]Jean-Luc Daval, "La relation comme interrogation," in *Relation et relation* (Liège: Yellow Now, 1981), p. 102.

[4]See Umberto Eco, *Opera aperta* (Milan: Bompiani, 1962). Eco has redefined his terms in the English-language edition: *The Role of the Reader: Explorations in the Semiotics of Texts* (Bloomington: Indiana University Press, 1979).

Language Processes and the Metalinguistic Puzzle

Daniel Simeoni

INTRODUCTION

Experts in animal physiology together with artificial intelligence (AI) researchers tell us that the operational complexity of human vision is incommensurate with that of lower-level species in the evolutionary scale. Vision being essentially a utilitarian function related to survival, it follows, for example, that the fly's visual apparatus differs from ours in terms of adaptability to the environment, inasmuch as a fly's expectations from the world are noticeably lower than humans'. In like manner, the

I am grateful to Gerald Mead for his competent support and friendly help, sometimes provided at the shortest notice. I wish also to express my thanks to Bonnie Osborne for her patience and good humor. As usual, Adrienne Chambon provided much of the inspiration and, in one case at least, specific reference to academic fieldwork supporting my views: My debt to her cannot be overemphasized.

world as we see it contrasts radically with the reality experienced by other, less demanding species. It is hardly surprising, therefore, that the problems faced by AI experts to account for the transformation of the human retina's bidimensional image into moving three-dimensional colored objects as we perceive them are still an open challenge.

Similarly, matters pertaining to the phylogenetics of language remain highly conjectural questions; however, recent developments in the field of computational semantics and, more generally, in cognitive science research yield substance to the assumption that the emergence of language in humans may have been coeval with that of our "advanced vision." Judging by the interest and excitement generated by several recent publications, there is little doubt that the interaction and common properties of the two systems of representations are as significant as their apparent differences.

Many recently devised linguistic models explicitly pose an operational analogy between language and vision in terms of their derived structures or underlying processes. To the outsider, this analogy may seem like just another innocuous idiosyncrasy. Linguists, one sometimes hears, are fond of devising all-encompassing, definitive theories, and the literature teems with formal models, hardly conceived, quickly discarded, sometimes barely debated for want of peer recognition. Yet seen from a longer perspective, once the dust settles, the current strategic shifts in the theory of language may prove to be nothing less than the building stones of a minor revolution in the state of the art: the return of the supremacy of representation over sign-induced preoccupations.

This chapter begins with a quick survey of what could be called the Gestalt takeover of modern linguistics; moves on to the implications of this paradigmatic change for the definition of an adequate metalinguistic tool; provides methodological guidelines for implementation, using explicit examples taken from current theories of language on both sides of the Atlantic; and, finally, sketches the broad lines of a research agenda involving the participation of experts of both verbal and visual persuasions.

PART I: CURRENT CRACKS IN THE LINGUISTIC SPHERE

Traditionally, contemporary theoretical linguistics[1] has not had much to do with the hypothetical constructs of less "hard" cognitive sciences. Most of its proponents—including those who stated that their research was a subfield of psychology,[2] explicitly setting their scope of inquiry within the wider mold of biological studies[3]—seem to have been content with giving the matter quick lip service, thereafter hastily resuming their quest for the definitive theory of language in terms of universals,[4] truth values,[5] and other traditional notions such as belief contents.

Stated simply, the dominant paradigm in the linguistic sciences to date has been the world of logic. Whether in semantics, in syntax, or in phonology (determining such things as the smallest significant unit in the speech flow of a language), the approach has been consistently analytical (dividing the data into discrete units), structural (finding out which constructs could be cohered into stable "levels" of treatment), and, in short, operationally exclusive (that is, based on the supposedly all-purpose either-or operator). The reasons for this epistemological approach remain unclear and need not be reviewed here; suffice it to say that these logico-mathematical foundations have had such an impact on research, in general, that each time their validity was challenged, serious rifts ensued within the ranks of the linguistic profession.

The advantages of the traditional approach are not to be sneered at; not surprisingly, the serializing or "slicing out" of utterances into primary units allowed for the emergence of a host of categories and classes grouped under specific headings, which, in turn, gave way to a proliferation of grammars and a classification of the world's languages. Indeed, the work accomplished by linguists along those lines bears resemblance to that of the Russian chemist Mendeleev, who ordered the flow of matter and classified it relative to its smallest, irreducible elements (Figure 1) or, for that matter, to Aristotle's Analytica.[6] This tradition is no historical curiosity; this very same approach of language reality has so far informed most AI attempts to devise parsing algorithms.[7] The exclusive order translates nicely into binary combinatorial architecture.

The problem with this approach is that its power of treatment is tragically limited. In particular, it ignores a number of crucial phenomena without whose integration natural languages cannot, I believe, be accounted for.[8]

Language as it is spoken in everyday circumstances is rarely grammatical. Even though the general picture is clear, it is often misunderstood and, at best, partially understood. The meanings it conveys are essentially shifting entities, and it is simply not proved that an analysis of the speech flow into discrete units at any level clarifies what is involved in the processes of speaking or understanding a language. In addition, the logico-analytical approach is inadequate when it comes to explaining illusions of meaning (just as there are visual illusions), how comprehension of a distorted, sometimes self-contradictory statement is possible, or how an argument develops.

These phenomena are self-evident; they form an integral part of a native speaker/hearer's daily experience of language. However, to un-

Group 0	I a	I b	II a	II b	III a	III b	IV a	IV b	V a	V b	VI a	VI b	VII a	VII b	VIII
	H 1														
He 2	Li 3		Be 4		B 5			C 6		N 7		O 8		F 9	
Ne 10	Na 11		Mg 12		Al 13			Si 14		P 15		S 16		Cl 17	
Ar 18	K 19	Cu 29	Ca 20	Zn 30	Sc 21	Ga 31	Ti 22	Ge 32	V 23	As 33	Cr 24	Se 34	Mo 25	Br 35	Fe 26, Co 27, Ni 28
Kr 36	Rb 37	Ag 47	Sr 38	Cd 48	Y 39	In 49	Zr 40	Sn 50	Nb 41	Sb 51	Mo 42	Te 52	Tc 43	I 53	Ru 44, Rh 45, Pd 46
Xe 54	Cs 55	Au 79	Ba 56	Hg 80	57-71*	Tl 81	H' 72	Pb 82	Ta 73	Br 83	W 74	Po 84	Re 75	At 85	Os 76, Ir 77, Pt 78
Rn, 86	Fr 67		Ra 85		Ac 89		Th 90		Pa 91		U 92		Np 93		Pu 94, Am 95, Cm 96

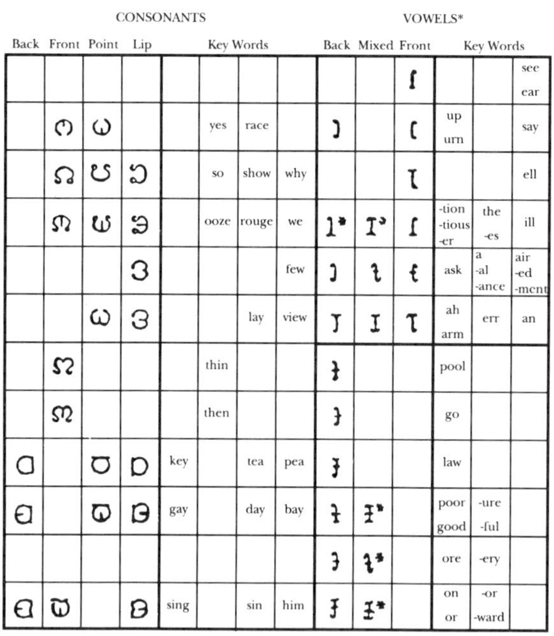

FIGURE 1. Short-period form of periodic system of elements and Bell's "Table of English elements, showing their position in the universal alphabet" (Encyclopaedia Britannica and A. H. Stewart, 1976).

derstand fully the implications of turning them into central objects of study, the following ought to be recalled: Up to the last decade or so,[9] these phenomena were relegated out of the theoretical mainstream under the joint authority of no less major figures than Saussure and Chomsky. Only recently have more researchers worked on more integrative models, thereby trying to encapsulate what had been consigned to a theoretical back shelf, under the headings of "parole" and "performance."[10] Within the framework of this surreptitious, almost clandestine, Copernican revision (in a reversed sense)—a radical repositioning of the linguistic object of study combined with the conviction that the logico-analytical paradigm by itself was not sufficient—here and there "new tunes began to be piped." These "tunes" hinted that to understand better the underlying machinery at work in the activity of language, a good idea might be to begin with novel hypotheses, on the basis of such unorthodox constructs as mental images[11] or even visual precepts.[12] That such referential analogies were at the inception of new linguistic models is no insignificant thing; Aristotle ranked analogy among the classic figures of logic, on a par with today's widely accepted modes of deduction and inference.[13] Only the advent of modern science, with its emphasis on axiomatic necessity, has made analogy lose its theoretical power and, along with it, much of its respectability as a methodological tool. Recent contributions in the history of scientific discoveries have confirmed that things were not that naturally clear-cut.[14]

Although originating from the same frustration, these attempts were conducted and developed independently of one another. Their authors make no claim of being part of the same movement; their theses present profound dissimilarities; occasionally they even contradict one another. Yet it would be hard to ignore that on crucial issues their research was based on the same postulates and, at times, resulted in astonishingly common findings.

At the heart of the theory, they pose selective representation; meaning does not exist out there, waiting to be accessed, and language is not a code whereby the world around is deciphered but a function making the things we talk about meaningful. Reality, as we have access to it, is a creation of our language, much as it is a creation of our senses. From this point of view, nothing distinguishes the words we use from our precepts or, for that matter, the sense we make of a sentence written or enunciated in our native languages from a psychotic's hallucinations.[15] Conceptual structure is at least partly related to visual imagery, and reality is speaker-, hearer-, and viewer-centered, *not* object-centered.

The *francs-tireurs* of modern linguistics do not reject squarely the views of the traditional school; when they do not ignore its approach, they mold it within the framework of the new models. They do not dispute the value of such well-attested concepts and findings as those of the double articulation of language,[16] the arbitrariness of the sign,[17] and the reduction of language to an orderly sum of discrete units. They simply say that these distinctions are very wide of the mark if what is being looked for is an account of language productions in all their diversity and creativity. Creativity, in particular, cannot be disposed of in the theory by reducing it to the speaker's genetically preordained faculty of producing hitherto-unheard-of statements.[18] Creativity must also account for what Henri Frei[19] characterizes as "*l'expressivité dans la langue*," that is, the oddity that there is something more about language than the sum of its individual components. In that sense, there is a Gestalt approach of things linguistic, in contrast to the prevailing Cartesian mode of the past few decades.

PART II: A METALANGUAGE PUZZLE

One of the most difficult problems faced by the proponents of the Gestalt approach is that of a suitable metalanguage. It may also be the major obstacle to its supporters being considered as members of a movement; as long as no common treatment medium can be agreed upon, their often extremely complex systems will be perceived as isolated "pot shots." The traditional school has long made use of mathematical expressions, from equations of first-order logic to fuzzy-set applications. This has always been particularly true of syntactic studies,[20] but even more recent models attempting to integrate unashamedly semantic components[21] show that the logico-analytical paradigm is still very much in favor. A reason might be that, although few linguists today adhere blindly to the Cartesian dogma, a coherent substitute for the traditional scientific tools of logic has not been found. This chapter unequivocally argues that the quest for a common working language must now take priority over disparate attempts, and it suggests how this can be accomplished.

Gestalt linguists today face basic problems (Figure 2). How can one account for the following?

1. A given acoustic/graphic form in any one of the world's natural languages translates effortlessly into a conceptual whole.[22]
2. A given conceptual whole can translate (although sometimes with much greater difficulty!) into any one of the world's natural languages.
3. Neither process is a "bijection" (a surface form will translate into several more or less distinct meaningful forms, and a conceptual whole will translate into several plausible surface forms).

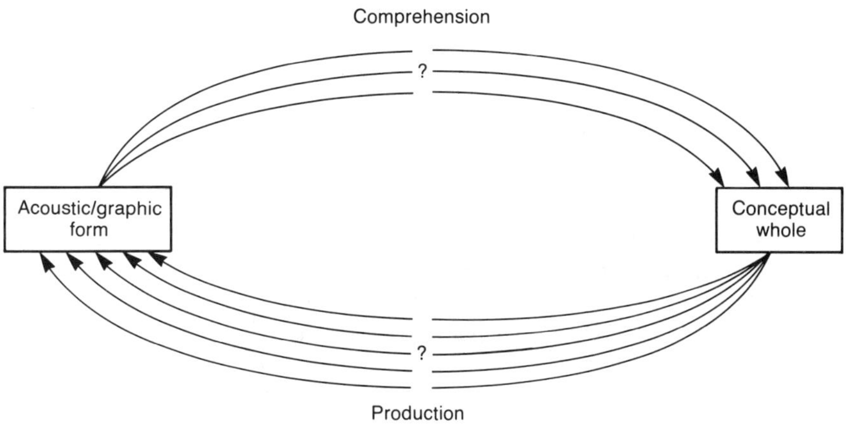

FIGURE 2. Basic language operation.

4. Consequently, the two processes are not reducible to each other (or, as a logician would say, the operation is not symmetrical).

Accounting for these four phenomena would require

- A suitable metalanguage (Since classical logic only tells us that the fundamental linguistic relation is *not* a bijection and that it is *not* symmetrical, then what *is* it?)
- Finding an algorithmic (as opposed to analog or figurative) formula whereby the transformation might be described and controllable
- Ideally, for the sake of scientific characterization, both the metalanguage and its supporting algorithm(s) being refutable

One point should be vigorously emphasized here: Following a non-Cartesian approach does not in itself preclude the use of logic. Only the level of application must be higher.

The plight of linguistics has long been that it is the only science whose objects of study—verbal artifacts—are also its tools of treatment, with the additional curse that all these utterances are simultaneously engaged in a state of radical indefiniteness. Ad hoc definitions, however precise and tuned conceptually, are of no avail here. Although unpleasant to admit, the linguist's metaexplanations are subject to the same limitations as ordinary, everyday language. One simply has to think of the fortune of terms such as *sign, performance,* and *representation* (the list is virtually unlimited) to be seized by the all-pervasive ambiguity of linguistic symbols. The *improbable* linguist who would make exclusive use of such terms to elaborate a theory of language ought not to be surprised if readers show a propensity to misunderstand his point. Every teacher knows that such confusions are inevitable, given the potential polysemy of natural language statements.

More often than not, this lack of specificity of ordinary language is a harmless thing.[23] Depending on one's sphere of activity, it is said to warp communication or to facilitate it; no serious claim is made that it prevents communication. This paucity of semantic rigor may also be a trade-off for the flexibility allowed by usage. In short, only the linguist's formidable pretension to scientificity creates a problem.

Luckily, the inability to reduce "performance" acts to straightforward logical analysis has no mandatory counterpart in the case of an appropriate metalanguage. In other words, no such polarity need affect the chosen metalinguistic forms, unless the metalanguage in question is expressed in a natural language. The trick is simply to find a paradoxical metalanguage that will stick to the fuzziness and lability of natural languages[24] and, at the same time, conform to the requirements of ordinary scientific usage. Admittedly, this is a tight rope to walk, but there is growing evidence in current literature that it can be done. Besides, it seems to be the indispensable price to pay to avoid falling into the traps of irrelevance or mere "commentary."[25] Certainly, both ordinary language and the language of logico-mathematical relations are necessary theoretical adjuvants (the former at the level of interpretation, the latter as a criterion of scientific validity). However, to expect either of them to ever contribute an adequate metalinguistic tool is an illusory goal. Possible clues to the solution of this puzzle are provided in the next section.

PART III: THE THEORETICAL STAGECOACH, FROM THE EAST COAST TO EUROPE AND CALIFORNIA

Refocusing the problem along those lines invites comparison with the tasks set by AI experts for other well-recognized objects of study, such as form and movement recognition or sound perception. The key word is simulation, but the basic idea is the same: How can one account for these amazingly efficient, unwillful processes in a way that reproduces the actual biological operations at play as faithfully as possible?[26]

Because many AI experts assume that perceptual operations are strictly constrained, the most stress is laid on the combinatorial rules that permit so-called "primitives" to transform gradually into meaningful forms. A typical question is, how can such rudimentary building blocks as the blobs and bars that form on the human retina combine with the regularity needed to produce our environmental images? Because of this way of phrasing the problem, the emerging *structures* that cohere at every intermediate level get the limelight, rather than the operating *processes*. Strictly speaking, the perceptual model currently in vogue is a hybrid; its proponents make extensive use of Gestalt principles, even figures, to define the terms of their problem, but they opt for a logico-analytical line of reasoning in the course of solving it. Even though the results of the transformations are widely acknowledged to be phenomenological wholes (whether they be images, sounds, or sensations), the models proposed are clearly the upshots of a crosscutting, multitiered thinking process. The so-called "modularity of the mind" revival is the most recent example of this classificatory mode.[27]

A way of accounting for the apparent success of this approach in perceptual studies could be that these problems are genuinely simpler than those posed by language. The primary linguistic relation (see Figure 2) is a two-way process. Moreover, the diversity of natural languages requires at least one supplementary level of treatment. By these standards, the perceptual model looks fairly straight, and the restrictions on an appropriate metalinguistic tool simply may not apply in its case. Whatever the reason, the metalanguage currently in use in perception studies is heavily logico-mathematical.[28] The unstated argument for this approach would run somewhat as follows: The tighter the processing constraints, the easier it is to use a computational model. A corollary is that linguistic processes are too loose; they allow for much too much leeway in real acts of language to tolerate such rigorous categorizations as incorporated by logic.

Despite these restrictive provisions, attempts have been made recently to map a theory of language directly onto a perceptual model of representation. Ray Jackendoff's formal apparatus[29] for describing the processes underlying language production and comprehension is of particular interest, as it connects with counterparts in the five senses and closely parallels David Marr's computational theory of visual representation.[30] However, the most original aspect of Jackendoff's thesis lies in the way he attaches the semantic component of verbal occurrences to the conceptual value of visual images. In this linguistic interpretation of the perceptual model, Marr's 3D model of human vision (in Jackendoff's words, the locus of "central representation") becomes the hub around which both visual and verbal comprehension as well as conceptual formulation are said to revolve (see Figure 3).[31]

Best of all, Jackendoff's suggestion that conceptual wholes may have visual components, both in language production (before translation into graphic/acoustic form) and in comprehension (after it) brings him just short of questioning the validity of analytico-semantic tools in linguistics;

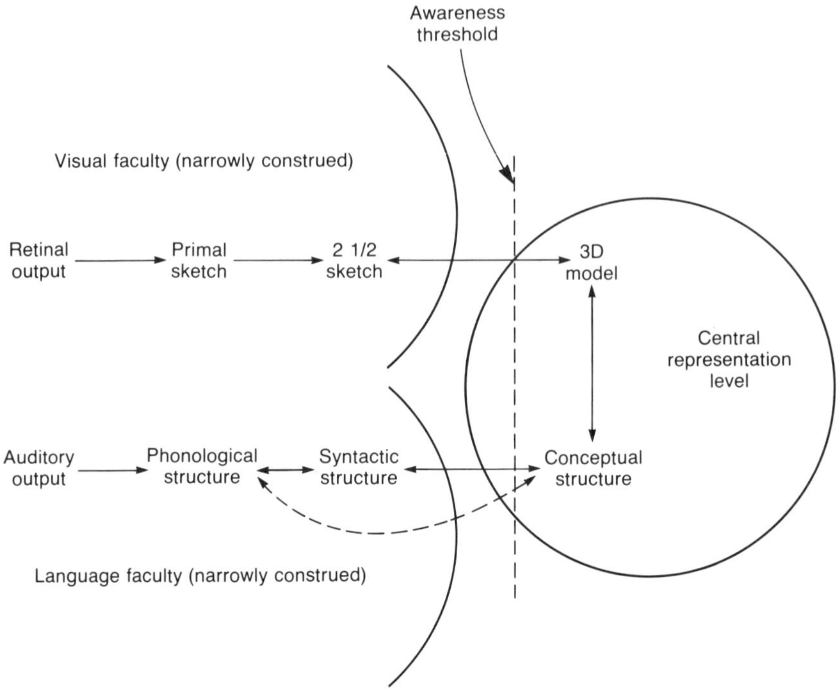

FIGURE 3. Articulation of the visual and language faculties according to R. Jackendoff (adapted from CCM, p. 248).

the problem of speech relations in general remains unmentioned but word meanings, for example, clearly are susceptible to nonanalytical modeling in his view: "Although the formal niceties of such word meanings are yet to be worked out, I think it is possible to see the germ of an important descriptive advance here. By using linkages of 3D models with conceptual structure, one can begin to circumvent the limitations of the purely algebraic systems to which semantics has been largely confined and at the same time begin to see how language can make contact with the world as perceived."[32]

However, a true adept of the perceptual model of representation, Jackendoff hastily reverts to the crosscutting, exclusionary mode of reasoning once the terms of the problem have been defined. He strongly argues for the primacy of the structure-based approach over the processing strategy; the last words of his book are yet another, final attempt to legitimize his choice: "My suspicion, anyway, is that the actual function of consciousness can be revealed only by understanding its *structure first*."[33]

Jackendoff's choice of a structure-first interpretation should not be much of a surprise because he is also a generativist, trained to linguistics through the most classic school of syntactic studies. Indeed, if there is anything surprising about his theory, it is his rephrasing of the linguistic enigma from the point of view of an experiential subject. Jackendoff's thesis is exemplary of the epistemological shifts that are currently shaking linguistic circles. A computational study it certainly is; but the object of study and, correlatively, the angle of approach are unequivocally phenomenological. That such a questioning of the very nature of linguistic research should have matured and arisen within the ranks of the supporters of the dominant paradigm is proof of how great the need is to supplement the austerity of this approach.

A more radical perspective, though, is provided by other, self-avowed outsiders. Although they form distinct systems whose roots can be traced to equally distinct traditions, both Ronald W. Langacker's and Antoine Culioli's theories of linguistic processes are complex attempts to formalize the open, unstable, yet somehow deeply regulated speech flows that make up language experience.[34] Central to their theses is the same idea that "much in language is a matter of degree,"[35] making it necessary for the linguist to devise specific metalinguistic tools to capture the constantly moving, holistic reality of verbal artifacts. How, for example, can we account for the difference in meaning between *raw* and *cooked*?[36] Everybody knows when something is cooked (the same is true, of course, of all the intermediate phases, such as rare, medium rare, and so forth) and yet a precise *dividing line* between any two contiguous terms is impossible to determine. To deal with such intrinsic shifts of meaning, Culioli has devised the notion of *gradient* and Langacker that of *sequential scanning*. The difference in terminology matters less here than the common determination to tackle the problem head on or, in Langacker's own words, the recognition that "The analytical convenience of assuming that everything in language is discrete encourages linguists to develop discrete formalizations, resulting in general neglect for the many aspects of linguistic structure not amenable to such treatment.[37]

Other equally uncommon tools of treatment[38] include terms such as: *notional domains, boundaries, course, attractors, extraction* (Culioli); *primary domains, schematic network, autonomy/dependence asymmetry, focal adjustment* (Langacker). Both linguists define language as "meaningful representational activity" (Culioli) or "symbolic in nature" (Langacker), denying syntax any radically autonomous character. Their primary sources of inspiration are in cognitive research and anthropology. (They both cite Eleanor Rosch's works on prototypes,[39] making comparable use of this concept in the insights they offer on "schematic networks" and "notional domains.")

More than perception, *imagery* lies at the heart of treating language as a mode of representation. The term itself is glossed by Langacker as "the ability to construe a situation in alternate ways for purposes of thought and expression, e.g., by effecting various types of focal adjustments."[40]

Now even a cursory look at the literature on imagery reveals that the terms of the problem are no different from those in use in the field of perceptual vision.[41] My interpretation is that the new paradigm being developed in linguistics is based on certain analogical relations common to linguistic operations, visual processes, and, more generally, cognitive behavior and strategies.[42] The new conceptual tools fashioned by language theoreticians leave no doubt as to the choice of this orientation. The only remaining uncertainty concerns the implied methodology itself. Will research focus—as seems to be the case at present under the influence of AI models—on the structures that inform these operations? (Whether provision is made within such a framework for an autonomous syntactic element is beside the point.) Or will it concentrate on the underlying processes themselves, in which case, utmost attention should be paid to the nature of the metalinguistic treatment?

PART IV: AN ISSUE OF DESIGN

The improbable linguist mentioned in Part II does not exist; no current linguist would restrict his metaexplanations to the use of natural language statements. Opening a few random pages in the current literature shows that verbal forms are only one component in a full-fledged array of metalinguistic instruments. This array ranges from plain conventional symbols to icons and includes a variety of indices, sets, brackets, typographical subtleties, diagrams, and sketches (see Figure 4). This chapter definitely claims that these representations are not just formal niceties. They are forceful attempts to get around the ambiguities and the general fuzziness of linguistic symbols, the shortcuts meant to clarify an idea, the ways of buttonholing readers when the going gets a little too tough, and occasionally, the ways of circumventing the difficulty of translation.[43] In more ways than one, readers are thus put in the picture.

Such visual supplementing of the propositional mode is inevitable. At a certain level of abstraction in the theory, dramatic effects have sometimes been observed of linguists resorting to physical gestures in the middle of professional speeches to get their messages across. By their power of suggestion, these visual metaphors uphold the intuition that, *from a formal point of view, the sequential, linear concatenation of linguistic symbols characteristic of natural language artifacts is literally substantiated by the analogical, figurative, nonalgorithmic order of imagery*. Roughly speaking,

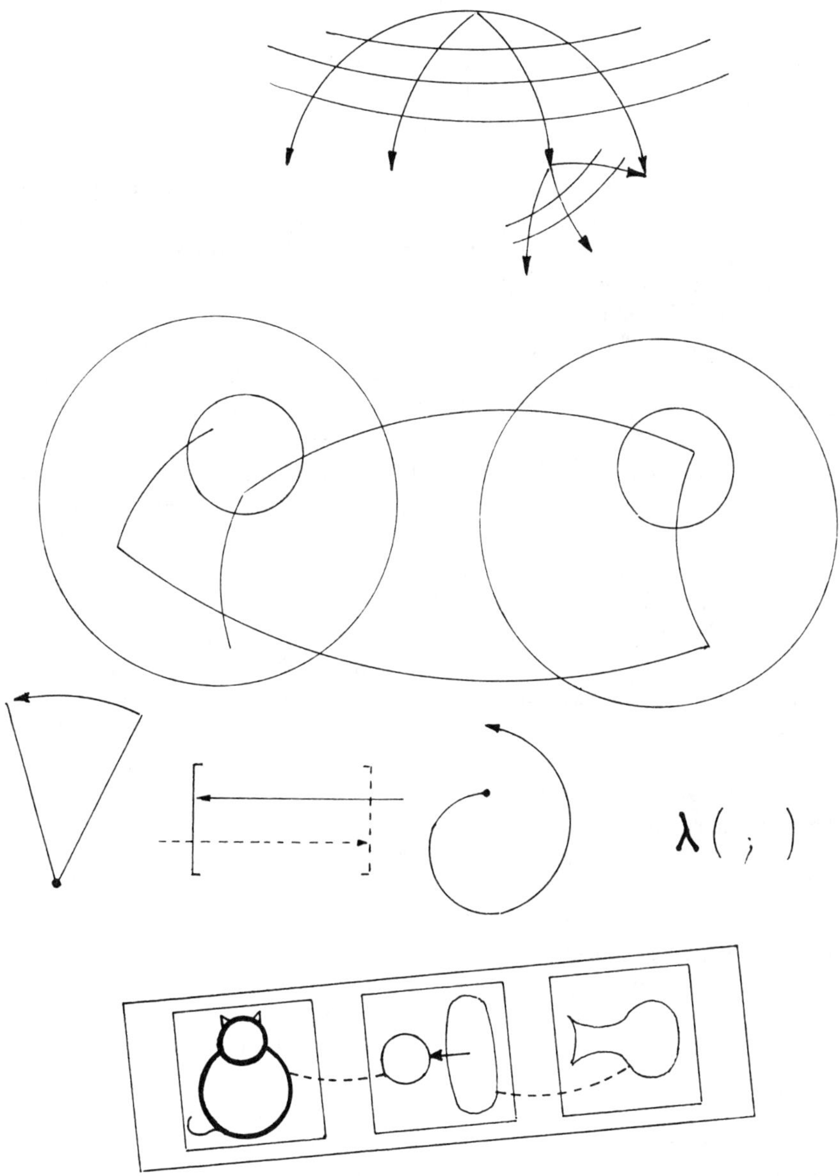

FIGURE 4. A few examples of metapictograms currently in use.

the interconnection works somewhat as follows: Conceptual wholes form on one end, and somewhere down the line, visual imaging helps bridge the gap as a prerequisite to language actualization. Far from being a conquetterie, *visual representation* is an integral part of the processes underlying speech production and comprehension. Proverbial psychological reality alone should therefore be sufficient backup for integrating the visual mode fully to the theory. I believe that laying the groundwork for a coherent metalanguage along these lines should be the primary goal of both language theoreticians and AI experts, in the spirit of the pioneering work accomplished by the aforementioned researchers. The following remarks describe possible pathways for future development.

Scientificity should be the major criterion; approximation is as much to be feared from figural treatment as it is from verbal explanations. Indeed, the risk is likely to be vagueness, let alone utter nonsense, if the usual precautions are not taken (see above-mentioned requirements, Part II). Examples of such confusion can be found in earlier, poorly conceptualized attempts.[44] The scientific status of methodology in the human sciences in general has been clearly defined by Bresson; strict logico-analytical treatment should be confined to the metatheoretical level, thus guaranteeing the validity of the method chosen.[45] This kind of approach is exactly what has been argued for in this chapter. Applying the structures of classical logic directly to the empirical data provided by natural usage can only result in spurious scientific rigor, "reducing the issues to artificial phenomena." Linguistics as a science is no different in this respect from other attempts to formalize living, organic systems. Only the supporting algorithms should be accountable to the laws of logic, not the metarepresentation as such, let alone the data under scrutiny!

Comparison of iconic and linguistic symbolic systems is not a new field of inquiry. Experiments[46] have long established that teaching the technique of knot-making verbally is impossible without the complement of visual representations. (To have an idea of what is at stake, imagine trying to learn knitting by telephone.) Yet in most cases, existing literature finds arguments there to contrast and oppose the two systems, usually downplaying the role of visual symbols by comparison with the interpreting power of both natural language *and* logico-mathematical systems. Typical of this line of reasoning is Bresson's observation that the latter systems share, "because of their general syntactic organization, a universal capacity (at least that we may think of as such) for treating mental products [whereas figures and images] can only express states and as such, are inadequate to communicate."[47]

Similarly, at the other pole of the semiological field, it is often claimed that speech and image belong, in essence, to two radically distinct spheres of study, as the linguistic and visual variables that inform them are fundamentally divergent. Attempts at drawing them nearer are usually hastily dismissed, on the grounds that "Symbolism demonstrates that anything can mean anything through analogy; this is usually of little value in efficient visual communication."[48] Such unanimity in protecting one's preserve can only lead to one conclusion: Old discriminations too die hard.

Again, my impression is that this distribution of roles between the algorithmic and the analogical may be a consequence of the approach favored by traditional semiologists rather than a truly inherent clear-cut division. For example, it seems to exclude indices (such as arrows and ordered schemata) from the category of figural symbols. More radically, it does not take account of the time dimensions in its analysis of the two systems. If communication works so well with words and algebraic symbols, their processing features are the cause, not some inherent characteristic in their design. Just introduce the processing dimension in the iconic system and the outcome will be just as efficient in terms of communication as with any other arbitrary construct. Conversely, extract a structural crosscut from a natural language expression (or from a mathematical formula, that is, by leaving out the operators) and no communication, that is, no sense, will take place. To be workable, a visual metalanguage would likely require using the whole array of known figural resources. There is no *a priori* reason to exclude any technique. Indeed, there is every reason to look for techniques that incorporate several types of representation: iconic, index-like, spatial, and so forth.

That figural representation should be a crucial mode of treatment for "the *activity* of language as we apprehend it through the diversity of natural languages"[49] is not an entirely new thing. The expressive power of figures and schemata has frequently been noted. The reason why the gauntlet has never been picked up is to be found in the quasi-unanimous refusal by linguists, cognitive scientists, and semiologists alike to assign *processing power* to these representations. Although a number of precursors must be paid tribute for taking steps described in this chapter, those efforts do not seem to have resulted in much more than 3D figuration; the main obstacle has remained unchanged.[50] With the recent development of computer graphics, the time has come at last to reconsider the bases for such reticence and to embark, once more, on uncharted territories.

On every account, computer graphics "take the shine out" of other metarepresentational tools; because the graphic elements can be made to move and merge, they could—ideally—adhere to the plasticity of natural language statements; they could also reflect the inherent fuzziness of linguistic symbols, express illusions of meaning, illustrate ambiguities and uncertainties, and, in short, provide as close a description of "performance" acts as could be wished. Because such graphics combine several modes, their power of treatment would be unequaled. Above all, because they are made of nothing but numbers, their algorithmic accountability would be given *per se*. As to the artificiality of their images, perhaps the best comment is that so-called "natural" languages themselves are not so natural as would seem at first sight; even one's own mother tongue must be learned. Surface actualizations, whether graphically or acoustically informed, are no more natural than the bidimensional images that form on the human retina or, for that matter, than any information-bearing device. *Usage* only is natural and, in this respect, nothing will detract from computer graphics representativeness. It is sometimes hinted that current graphic creations have no reality and that they are detached, abstracted representations, with no organic relation to living material; yet even in those cases when no analog can be clearly identified in the surrounding world, these images must, at the very least, be acknowledged as the real, tangible products of the mind(s) that designed them.[51] Making use of such images in the representation of natural language creativity would no doubt help make the genetic relationship clearer still.

CONCLUSION: BACK TO REPRESENTATIONAL BASICS?

This chapter is indicative of an itinerary. The technical complexity of the suggested path calls for transdisciplinary work. Contacts among teams of researchers in all related disciplines will have to be arranged, as tentative solutions emerge.

Perhaps the most exciting thing about this suggested paradigmatic revision is that under the garb of technological novelty, it reaches back to the very origins of human language. If it is not too rash to assume that long after the fly's perceptual system had stabilized, human language moved from visually shaped conceptual contents to oral expressive forms and from there to writing systems, would it not be about time to realize that language as an object of study for linguists has been mostly restricted to the tip of the iceberg, that is, the sequential, linear chain of conventional signs out of which we manage to extract significant patterns? This preoccupation, indeed, this obsession with the nature of the "sign" should now make room for a revival of the seminal concept of representation. By applying a visual theory of representation to the verbal artifacts that make up the experience of language, we may be able to relate (and thus get closer) to the original conceptual source from which everything developed.

Notes

[1]Theoretical linguistics: the "hardest" of all language sciences. Focus is set on the *internal rules* that determine the formation of language occurrences, in contrast with psycho-linguistics, socio-linguistics, neuro-linguistics, and so forth, in which stress is laid on the *interaction* of language with other, distinct fields of study.

[2]"Linguistics is simply that part of psychology that is concerned with one specific class of steady states, the cognitive structures that are employed in speaking and understanding." Noam Chomsky, *Reflections on Language* (New York: Pantheon, 1975), p. 160.

[3]"We study language as a first step to making claims about properties of the mental genotype as a whole, the initial state of the human mind." David Lightfoot, *The Language Lottery: Toward a Biology of Grammars* (Cambridge, Mass.: MIT Press, 1982), p. 203.

[4]The quest for so-called "universal grammar," on the basis of detailed observation of a particular language, is a central claim of the transformational-generativist school. Similarly, following intensive fieldwork by such linguists as Joseph H. Greenberg, attempts have been made to determine universal language forms, that is, structural constants common to all known languages. See Joseph H. Greenberg, ed., Charles A. Ferguson and Edith A. Moravcsik, ass. eds., *Universals of Human Language*, 4 volumes (Stanford, Calif.: Stanford University Press, 1978).

[5]For an anthology of views focusing on the formal conditions of truth and falsehood and their relevance to the various processes whereby competence manifests itself in language, see J. M. E. Moravcsik, ed., *Logic and Philosophy for Linguists* (The Hague and Paris: Mouton, 1974), part IV, pp. 299–343.

[6]This criss-crossing of the reality observed is supposed to exhaust *all possible* combinations of elements, hence its "predictive" value. Hence also the claim sometimes heard from generativists that their linguistic theory is "too strong . . . to the extent that it provides a principled basis for selecting a descriptively adequate grammar for each language acquisition." Noam Chomsky in Herman Parret, *Discussing Language* (The Hague and Paris: Mouton, 1974), p. 40.

[7]With a few notable exceptions, such as Hofstadter's introduction of conceptual "slippage" or "slippability." See, for example, Douglas R. Hofstadter, *Metamagical Themas: Questing for the Essence of Mind and Pattern* (New York: Basic Books, 1985), pp. 173–296.

[8]Natural language: any of the human languages characterized by speech forms or a writing system (excluding sign language, pictograms, and so on).

[9]As often, when one major school is being replaced by another, overlapping occurs; one such precursor was Emile Benveniste, today considered to be the founder of the French "théorie de l'énonciation." See, for example, Emile Benveniste, "L'appareil formel de l'énonciation," *Langages* (Paris: Larousse, 1970). (Understandably, linguists are particular about the terms they coin in order to define their hypothetical constructs, and "énonciation" has been a consistent snag for translators for a long time now. My own version of the concept is simply "verbalization," as indeed the term has been used in much the same sense, glossed as "the process of producing an utterance," by Wallace L. Chafe in, for example, Parret, *Discussing Language*, p. 10.

[10]The overwhelming majority of the linguists who worked on real-life, contextualized occurrences were at that time psycho-linguists, socio-linguists, and so on, *not* language theoreticians as defined in note 1 above.

[11]I am very much aware that the terms "imagery" and "mental spaces," both central in Langacker's and Fauconnier's theories, are to be construed in specific ways and that they should not be confused as such with any vague notion of "physical images in one's head," or even be taken as *implying* a necessary relation between meaning and visual imagery. For unequivocal examples of how wary the two authors are of such confusions

arising in their readers' minds, see, for example, Ronald W. Langacker, *Foundations of Cognitive Grammar* (Stanford, Calif.: Stanford University Press, 1987), pp. 110–13 or Gilles Fauconnier, *Mental Spaces: Aspects of Meaning Construction in Natural Language* (Cambridge, Mass.: MIT Press, 1985), p. 152. In both cases, this imagery is defined as an "ability," nothing more. I will be just as cautious in indicating that any further claim made on the basis of these two linguists' elaborations is only my own speculation.

[12]Ray Jackendoff, *Consciousness and the Computational Mind* (Cambridge, Mass.: MIT Press, 1987).

[13]My reading here is that, aware as Aristotle was of the lesser "logical" power of analogy—either defined as "proportional metaphor" in *The Rhetoric*, III, chapter 5, 1407a, 14–15 or, more extensively, in *Poetics*, 21, 1457b—relative to the more traditional figures of syllogism and induction, he never inclined to dismiss it as inferior; indeed he made extensive use of all its forms, on one occasion at least letting the cat out: "It is the one thing that cannot be learnt from others; and it is also a sign of genius, since a good metaphor implies an intuitive perception of the similarity in dissimilars" (*Poetics*, 22, 1459a, 6–8).

[14]"In summary, when scientists hold a theory, they hold a particular mode of imagery as well." Arthur I. Miller, *Imagery in Scientific Thought* (Cambridge, Mass.: MIT Press, 1986), p. 312. Naturally, the real difficulty arises as soon as one tries to untie the knots of this conceptual interplay. Which comes first, the propositional mode or the imagery?

[15]It is not easy to become aware that hearing or seeing (as in reading, for example) a verbal occurrence is not the same thing as understanding its meaning. Meaning can literally be said to be hallucinated. A perceptual parallel, even more provoking, is the one proposed by Hoffman in a way that is remarkably reminiscent of the philosophy of language type of concern: "If, as I suggest, such hallucinations are not an exception but the rule, and if they are in fact a necessary concomitant of visual perception, how can one justify one's faith in perception? How is it still possible that seeing is believing?" Donald D. Hoffman, "The Interpretation of Visual Illusions," *Scientific American* (December 1983).

[16]The duality of linguistic symbols whereby a finite set of primary insignificant units combine and result in an infinite array of significant occurrences. A lesser known fact is that the author of this—typically—theoretical construct, the "double articulation of language," also viewed it in an evolutionary perspective, as "a type of organization [that] seems to have imposed itself on human communities as best adapted to the needs and resources of man." André Martinet, *Elements of General Linguistics* (Chicago: University of Chicago Press, 1964), p. 26.

[17]The first principle governing the nature of the linguistic sign: "The link between signal and signification (i.e., between the graphic/acoustic form and its meaning) is arbitrary" (or conventional, or again "unmotivated," that is, "it has no natural connection in reality"). Ferdinand de Saussure, *Course in General Linguistics*, part 1, chapters 1 and 2 (London: G. Duckworth, 1983).

[18]"Having mastered a language, one is able to understand an indefinite number of expressions that are new to one's experience, that bear no simple physical resemblance and are in no simple way analogous to the expressions that constitute one's linguistic experience; and one is able, with greater or less facility, to produce such expressions on an appropriate occasion, despite their novelty, and independently of detectable stimulus configurations, and to be understood by others who share this still mysterious activity." Noam Chomsky, *Language and Mind* (New York: Harcourt, Brace, Jovanovich, 1972), p. 100.

[19]Henri Frei, *La grammaire des fautes* (Geneva: Slatkine reprints, 1982).

[20]See, for example, Edith A. Moravcsik and Jessica R. Wirth, eds., "Syntax and Semantics," *Current Approaches to Syntax*, 13 (Orlando, Fla.: Academic Press, 1980) for a sampling of fourteen different self-contained approaches of language structure.

[21]Jon Barwise and John Perry, *Situations and Attitudes* (Cambridge, Mass.: MIT Press, 1983).

[22]I have referred elsewhere to these conceptual wholes as "signifiés-images," in an attempt to emphasize even more radically the hypothesis I made of the fundamentally "visual" nature of language processes (Daniel Simeoni, "Images, Language," unpublished manuscript, June 1987, DRL, Université de Paris VII).

[23]Poets and other language lovers, from C. Baudelaire to J. Joyce and F. Ponge, would, of course, strongly disagree.

[24]Recent metaphors being worked out in the field of psychotherapy, where language remains a mandatory pathway, relate to significant results obtained by teaching the subject how to allow for greater "space" for his anxiety, against the "analytical" mode of reasoning, also insisting on the *dynamic process* of "focusing." This experiential approach would amount to freeing or loosening the thinking processes at work in language by increasing its "fuzziness" or "liability." Of all people, scientists may reject such an approach as least congenial to their working habits. See Eugene T. Gendlin, *Focusing* (New York: Bantam Books, 1981).

[25]Glossed as "a major procedure of control and limitation that discourse imposes on itself." Michel Foucault, *L'ordre du discours* (Paris: Gallimard, 1970). Unlike traditional "commentary," the ideal metalanguage would no longer "spell out, at the end of the chain, what had merely been silently enunciated over there." Indeed, as we envision it, it would not tell anything; it would simply *show* the way things are, ironically rejuvenating the much-forgotten etymology of the verb "to speak": to call attention to.

[26]Opinions vary among the authors themselves as to the "psychological reality" of recent AI or theoretical simulations of both language and perceptual processes. Certainly, research should strive for *some* degree of psychological acceptability in cognitive operations studies.

[27]Jerry A. Fodor, *The Modularity of the Mind: An Essay on Faculty Psychology* (Cambridge, Mass.: MIT Press, 1983). For a radical criticism of Fodor's thesis, see Georges Vignaux, "Semantic Domains and Cognitive Processes" (Semiotics and Artificial Intelligence, ACFAS Conference, University of Ottawa, May 1987).

[28]"The visual input to the horizontal control system of the fly is [thus], completely described by the two terms: $r(\psi)\ \Psi + D(\psi)$." David Marr, *Vision* (New York: Freeman, 1982), p. 33.

[29]Jackendoff, *Consciousness and the Computational Mind.*

[30]Marr, *Vision.*

[31]Jackendoff, *Consciousness and the Computational Mind*, p. 194.

[32]Jackendoff, *Consciousness and the Computational Mind*, p. 201.

[33]Jackendoff, *Consciousness and the Computational Mind*, p. 327.

[34]Quotations from Langacker are all taken from his *Foundations of Cognitive Grammar.* Quotations from Culioli are taken from various sources, in particular Hansjakob Seiler and Gunter Brettschneider, eds., "The Concept of Notional Domain," *Language Invariants and Mental Operations* (Tübingen: G. N. Verlag, 1985), as well as from his seminar (1986–1987).

[35]Langacker, *Foundations of Cognitive Grammar*, p. 14.

[36]Culioli, seminar, 1986.

[37]Langacker, *Foundations of Cognitive Grammar*, p. 45.

[38]Defining these terms clearly exceed the limits of this chapter; it should be clear from their "sound," though, that they have little to do with classical, logico-analytical tools.

[39]Eleanor Rosch, "Principles of Categorization," in Eleanor Rosch and Barbara B. Lloyd, *Cognition and Categorization* (Hillsdale, N.J.: Erlbaum, 1978), pp. 27–48.

[40]Langacker, *Foundations of Cognitive Grammar*, p. 490.

[41]"Experimental results . . . provide persuasive evidence that what we have called mental rotation is in a significant sense an analog simulation of an actual physical rotation." Roger N. Shepard and Lynn A. Cooper, *Mental Images and their Transformations* (Cambridge, Mass.: MIT Press, 1982), p. 186.

[42]See Georges Vignaux, "Discours, schéma, les actions du sujet," *L'activité argumentative: opérations cognitives et langagières* (Paris: Université de Paris VII, 1983), pp. 838–1107.

[43]"Translation is a particular case of paraphrase." Antoine Culioli, seminar, 1986. The metalinguistic objects reproduced in Figure 4 are borrowed from David M. Perlmutter, Gilles Fauconnier, Antoine Culioli, and Ronald W. Langacker.

[44]I. A. Richards, *Language through Pictures*, quoted by Herbert A. Simon, "What Is Visual Imagery? An Information Processing Interpretation," Lee W. Gregg, ed., *Cognition in Learning and Memory* (New York: Wiley, 1972), pp. 198–99.

[45]François Bresson, "Problèmes de la mathématisation en psychologie," *La mathématisation des doctrines informes* (Paris: Herman, 1972).

[46]Vignaux, *L'activité argumentative; opérations cognitives et langagières*, p. 722.

[47]François Bresson, "Les fonctions de représentation et de communication," *Psychologie* (Paris: Gallimard, 1987).

[48]Luc Vanmalderen, "Investigations into the Image," *Design Issues*, 2, 2 (1985), pp. 50–55.

[49]Culioli, seminar, 1986.

[50]Ann Harleman Stewart, *Graphic Representation of Models in Linguistic Theory* (Bloomington: Indiana University Press, 1976), pp. 136f.

[51]Philippe Quéau, *Éloge de la simulation* (Seyssel: Champ Vallon, 1986).

Man, Money, and Time: Logic of Credit— Logic of Modernity?

Jeff Kintzelé

> "You can't get anything fundamental for money; but all the fundamentals may be achieved if you get everything else for money." (Hans Blumenberg)[1]

INTRODUCTION: THE STOCK EXCHANGE TORMENT AND THE RELATIONSHIP BETWEEN MONEY AND TIME.

Two Mondays—described as being "black" under the circumstances—were enough to shake the foundations of a system that beforehand shone and fascinated by its intangibility, an extremely complex, sophisticated, abstract, immaterial system left only to the experts: the financial system. Some say, especially at this moment, that the system is fiction. This does not prevent the so-called unreality from becoming very real in the face of its disturbing consequences.

The common man rarely understands what is going on in the

financial world. His ignorance regarding the financial system—a monster of abstraction—is equal only to the stupor and panic that gripped the dignitaries of the financial temple on these famous Mondays. The commentaries that were to be heard shortly after, apart from despondency and defeat, were of considerable naiveté. Everyone has understood the mechanism now: "If one buys it goes up, if one sells it goes down" or "one buys when it goes up and one sells (or one has to sell) when it goes down." The principle is very simple but does not make the when and the why of a crash more comprehensible.

After the first shock of the 1987 stock market crash had subsided, the interpretation stayed within the limits of one general consensus. The crash was apparently foreseeable, at least in hindsight, and must be related to some simple causes: the indebtedness and budget deficit of the United States, the steep climb of the interest rate, the ambition and recklessness of the "golden boys," the West German Bundesbank's decision concerning the interest rate, and Secretary Baker's speech. Another important culprit, a symbol of modernity, is the computer. Computer technology indisputably contributed to the severe variations of exchange rates registered on that first Monday. Since the 1970s, stockbrokers have possessed programs that produce buy and sell orders automatically as soon as values reach a certain level. The excessive "sensitivity" of these computer programs made them react too quickly and too harshly, which amplified the stock market vibrations—an interesting case of *Eigendynamik* (proper dynamic). In short, we live above our means, and because of the specific interests of different countries, the worldwide harmony necessary in order to control the financial system has been disturbed.

All these explanations make sense but arouse curiosity even more. Why did the phenomenon grow to this size? Probably it is too early to analyze what happened in every detail and to predict what will happen in the future. Forecasts are disparate, ranging from the blackest pessimism to the greatest optimism. Without lucid analysis, the risk of succumbing to moralizing is great. Therefore, the event should incite us to become more cautious and try to understand why it happened.

The common person, who at first was confused, has now acquired a new vocabulary: arbitrage, index trading, leveraged buyout, and so on. This person also knows that a lot of people have lost large amounts of money, amounts that surpass the imagination. Billions of dollars went up in smoke in a few hours—money that never existed. But this paradox does not perturb the common person so much, who is more worried about a possible recession. The experts, in order to explain what happened, talk about living within our means, and this applies not only to the few who speculate on the stock market, but everyone. So, without really knowing what is happening, or why, the ordinary person knows that sooner or later he or she will have to carry the consequences. Isaac Joseph writes: "It is always the poor who get the wind in the face."[2] This reminds us painfully of a similar event that does not stop worrying our imaginations: the 1929 crash.[3] But simple comparison would be too easy: 1987 is not 1929 just because 1929 is present in everyone's mind.

The following ideas are not directly related to the torments of the stock market. We have taken that event as a starting point because above all its economic, political, and moral implications remind us of things that had almost been forgotten: for example, the fact that there are humans, with human reactions, beliefs, and worries behind this highly abstract and immaterial system. And, perhaps more important, the event has underlined our ignorance of an essential aspect of modern life: money.

The most common things are often the least understood. This is true of the money-time relationship, which is the real starting point for the ideas to be discussed.[4] The relationship between money and time is an intertwined one and can be summed up for many people in Benjamin Franklin's well-known phrase "Time is money." It represented the ideology or rather the state of mind of people during a historical period (or at least the dominant part of a period). And it is still valid today. The phrase does not, however, reflect reality unless you turn it around. It is a lie, as Max Horkheimer says, "that if you have time, you have money, because with time alone you cannot get money."[5] The unemployed, the out of work—a person who has plenty of time, but time from which he can draw no capital—would surely not contradict Horkheimer. Further, an excess of "free" time can even cause a loss of direction.[6] The fact that with money one also has power over other people's time justifies the turning around of Franklin's phrase. "Money is time" because it means projects.[7]

In economics one of the three classical functions of money is explicitly related to time: the value-reserve function. Money is "length of time": Its continuity is guaranteed. There are simultaneously conservation and anticipation. "One says that money is in and by itself an instrument that must keep its value," writes the economist Henri Guitton,[8] and he continues, "Money is the 'thing' that suggests in the present realities that do not yet exist. Money is an anticipation of value."

In the eyes of the sociologist, the interactions between money and time are even more numerous and subtle. Money, for example, is characterized by its eternal and continuous availability. Better: Money symbolizes the availability of merchandise and services. And by this it affects people's temporal consciousness. Due to the sociologist Georg Simmel, we have been able to show that money is essentially an accelerator of social processes. "There is no more striking symbol of the completely dynamic character of the world than that of money. . . . Money is nothing but the vehicle for a movement in which everything else that is not in motion is completely extinguished. It is, as it were, an *actus purus*."[9] On the other hand, money also moves the dimension of temporal horizons (past/present/future), a point that particularly interests me here.

So money is not only an economic factor but also a social one. Let us then leave the financial world to recover from its fall and to reflect on its own daily operations.

MONEY AND CREDIT IN EVERYDAY LIFE

Everyday life, notes Claude Javeau, is the "place where all meanings are created and perpetuated."[10] And money has become an irrefutable and unalterable reality in people's daily life, as Fernand Braudel correctly states: "Money has an implication in all economic and social situations and is in consequence a marvelous indicator. The entire human activity right down to the most humble levels can be judged by the way money runs, becomes out of breath, complicates itself, becomes absent."[11] Money mostly affects people's lives by its absence or abundance. Apart from this existential aspect, money influences human consciousness by the form it chooses to take.

Among all the small computer revolutions that touch daily life, the one concerning money is not the least important. We are, one author tells us, on the edge of the *age of electronic flux*,[12] the coming of a society *wiped clean of unsteady paper money and checks* is announced. We are finished with the *vulgar material aspect of payment*. Money is becoming invisible (even if this term is a little imprecise).

We are now well into the experience of credit cards and electronic money, the high point of mounting dematerialization and abstraction that money has been pursuing since its origins. The last limits (spatial and temporal) are collapsing. For the price of depersonalization, money becomes theoretically (for the happy few who have it) available everywhere and at all moments. Money is only a numbered piece of data stored in the famous chip's memory.

Simmel's reflections on the passage from coins to paper money, developed in his *Philosophie des Geldes*, which was published in 1900, deserve further development. He carefully paved the way: growing differentiation, rationalization, and individualization; repression of emotional attachments and preponderance of the intellect and calculation. Simmel also talks about a switch from culture to intellect. "This process might be called the growing spiritualization of money, since it is the essence of mental activity to bring unity out of diversity."[13] No one can refute this since the recent happenings. One must stress the fact that money is a driving force in society as well as a passively influenced part of it.

The computerization of money is not a revolution or mutation. There is no rupture: The process is continuous and while undergoing major changes it stays within the limits of the aforementioned logic. So the evolution, "the development from material money to credit money [,] is less radical than appears at first, because credit money has to be interpreted as the evolution, growing independence and isolation of those elements of credit which already exist in fact in material money."[14] Today money is rarely the only form of payment; a function of credit has become implicitly integrated.[15] This is, therefore, a good opportunity, in connection with the monetary phenomenon on one side and modernity on the other, to talk about credit.

BUYING ON CREDIT

We do not need liquid money any more, nor checkbooks; the credit card is enough. One signature and the desired thing is yours. This has many influences on our consumer habits. At a closer look, buying with credit cards is only one aspect of a vaster movement: consumer credits and the whole credit economy in general. Recently a significant climb in consumer credits has been noted in Europe. In France, for instance, a 43 percent increase was registered in 1986 and continued at the same rate in more recent years. The French, however, belong to the modest part of the class of people with "short-term debts." Only 3.6 percent of their salary is involved, compared to 7 or 8 percent for the English and 13 percent for the Germans, without mentioning the Americans. Probably things will get worse. One can buy everything with credit cards: cars, clothes, stereos, and even holidays. Here we have only mentioned what is classically called "consumer credit." One can go further into a larger category: personal credit.[16] This would allow us to include building loans, student loans, and so forth. This is the only way to know the extent to which an individual can create his whole existence on borrowed money and to show the real dimensions of living on credit.

The consumer-credit boom is due to the fact that the principal actors concerned feel they can benefit from it. In a period of stagnation or fall of buying power, households can keep up their standards of living. It is in the interest of the shopkeepers, of course, to stimulate the consumers' appetites; and the banks have discovered a new market (without even mentioning the spiky problems surrounding the creation of money, which is always a question of power).

However, indebtedness has taken on frightening proportions; headlines could shout: "A hundred thousand Belgians desperately indebted," "Forty percent of French households indebted." Many people are unable to pay back their debts and get caught in a vicious circle. They take out credit in order to pay back credits taken out beforehand. This situation is alarming, especially since we know only too well what the disastrous results of this mechanism are on a wider scale: the indebtedness of developing countries. Even if the credit to individuals in all its forms (payment by installment, personal loans, permanent credit, overdrawn accounts) is what we are mostly interested in here, we must not forget that this is a minor aspect in global terms. In terms of cost, this form of credit does not compare with other forms: credit to provinces or states, to towns, to firms, and to developing countries.

DIGRESSION ON THE CREDIT PHENOMENON BASED ON SIMMEL'S THOUGHTS

We have just noted that the coming of credit cards and the "take-off" of consumer credit and the economy of credit in general are of recent date. Actually the phenomenon has been in existence for a long time. It has a long story even though it did not always get much publicity. In the Middle Ages the development of credit was essential to the birth of commercial economics. "Unstinting credit was the great lubricant of the Commercial Revolution. It was altogether a novel phenomenon," notes Lopez. "The take-off of the following period was fueled not by a massive input of cash but by a closer collaboration of people using credit. . . . These modest contributions of the existing capital went a long way because credit enabled a small investment of hard cash to go to work simultaneously in more than one place."[17] The polemic around credit, or better around its acolyte "usury," constituted the "birth of capitalism," as Jacques Le Goff notes.[18]

Simmel starts by presenting credit as the extended form of money ("die Steigerung des Geldes"). The few considerations on the phenomenon are treated in his global reflections on money, especially when he analyzes the influence of money on life-style (considering predominantly the temporal aspects). Credit extends the series of conceptions still further and with greater awareness of their unrestricted breadth than does the intermediate instance of cash. The pivot of the relationship between creditor and debtor lies, as it were, outside the straight line of contact between them and is set at a farther distance from them: The individual's activity and transactions thereby gain the qualities of far-sightedness and enhanced symbolism. In that the bill of exchange or the concept of money debt in general represents the values of distant objects, they are condensed, as it were, in the bill just as the view over a spatial distance condenses the contents of the view by a perspectival shortening. And just as money places a distance between ourselves and objects and also brings them closer to us—thereby displaying its specific indifference in these contrasting effects—so too the instrument of credit has a dual relationship to our total assets."[19]

The logic of money is a logic of distance that reaches its highest point in credit: "die Steigerung des Geldes." Dieter Claessens has analyzed the consequences of this logic of distance in modern rationalism.[20] Distance permits, among other things, the growth of the possibility of anticipation. So we find in rationalism a kind of vectoral thought, a thinking by proportions and by consequence "formalization." This altogether means that individual content becomes replaceable. This possibility to replace contains the possibility to exchange. The concept of rationalism encourages an outlook of calculability about life, a calculator attitude.

Modern rationalism is more like a vision of the world in which the different contents have become undifferentiated, that is, means of equal value. As one has to give a notion of value, even just for pragmatic reasons, one can say that this vision of the world considers the world as unconcerned, unimportant, indifferent.

If credit is the extended form of money, then certain fundamental aspects of money on one hand and the consequences on life-style on the other hand become more apparent. Truly enough, certain aspects concerning the essence of money crystallize in the credit phenomenon. We might evoke, for example, the notion of trust that seems to be the cement of the monetary system. "If the embodiment of the action of exchange in a particular form is accomplished technically in such a way that each object is first exchanged for this, rather than directly for another object, the question arises: Upon closer examination, what is the corresponding behavior of people behind the objects?" Bruno Accarino considers this to be the fundamental theoretical question in the *Philosophie des Geldes*.[21] And Simmel answers: "The abstraction of the process of exchange from specific real exchanges, and its embodiment in a distinctive form [that is, money], can happen only if exchange has become something other than a private process between two individuals which is confined to individual actions."[22] The "soziale Gesammtheit," the entire community, has wedged itself between individuals. A "Zentralmacht" (central power) appears that creates and guarantees money with legal means. Paying with money then becomes a public rather than private obligation.

The worth of money lies in the trust one has in it. So modern society is a credit society. Belief and trust impregnate the fiber of society. A society that is made up of strangers needs trust and belief because trust, as Luhman notes, "is a factor that may reduce complexity."[23] Simmel goes a step further in considering trust as prerequisite for life in society; it prevents chaos and paralyzing fright. In our days (but probably in the past too) it is impossible to develop plans of action that take into account all factors, all possible outcomes. In order to reduce complexity, humans have used certain strategies: Rationalism is one of them; trust is an alternative. One can, however, suspect that trust is a ruse, a rationalization. "Thus, trust reduces complexity far more quickly, economically, and thoroughly than does prediction. Trust allows social interactions to proceed on a simple and confident basis where, in the absence of trust, the monstrous complexity posed by contingent futures would again return to paralyze action."[24]

Modern society is based on a reciprocal half-knowledge. The type of trust we are talking about here is, in David Lewis and Andrew Weigert's terms, routine or cognitive trust. It is in this way that our relative ignorance concerning money must be understood. People are satisfied if money plays the same role it has always played. So the institutionalization of money is largely due to the phenomenon of routinization and a tendency to make complexity into something commonplace ("banalization").[25]

In this chapter we cannot exhaustively examine all the guarantees that are required and conditions that must be fulfilled in order to be given credit. If we could, our analysis of cognitive or routine trust would likely be confirmed.

There is an evident connection between credit and social status. As a general rule, the monthly repayment of the borrowed sum must not be higher than one-third of the salary (after tax). The bank feels covered by this 33 percent rule. However, there are many uncertainties: the bank's investigation of its client's situation (which is already an intrusion into the individual's private life); protest; noting the summons to court; tallying unpaid checks (this last one is not official). However, the bank's information is not able to indicate the real situation: For example, the extent of an individual's personal debt or whether taxes have been paid for the past two or three years (a kind of credit given by the state) may not be known. In some countries there is no communication between banks either. In short, many unknown factors remain and the final decision is usually based on the banker's trust in his client.

Bankers often find themselves in a very ambiguous situation. They know about credit abuse and the fact that people often forget that credit means loaned money, but, on the other hand, they are not responsible for the moral conduct of people. It is their job to make deals: "Banking is business." They must take risks; otherwise their competitors will.

If things go wrong, it is always possible to limit damage by court appeal. In Europe this has only happened up to now in rare cases, but this does not mean a lot. In the case of Third World countries, credit was given based on the belief that a state cannot go bankrupt. Now we know that it is possible. Following the same kind of logic, the credit given to individuals is based on absolute trust in the legal system and the certitude that one can recuperate one's stake, an attitude that is very risky, especially when the "law of large numbers" takes over. On the whole, one must take into account the fact that the guarantees demanded by banks are only a half-assurance for them. The market is, however, covered by a fascinating game of insurance and reinsurance. In this way the banking system excels in pushing its sophistication further and further into unreality.

LOGICS OF CREDIT LIFE

Simmel showed us that credit has a "distancing effect" on peoples' lifestyles. We must now find out the consequences of credit on peoples' mentality. Everything points to the idea that the fundamental stake lies in the effect upon temporal consciousness.

Once again, we cannot avoid the analysis of the relationship between money and time. This relationship is, in fact, the basis of recourse to the trust mechanism from the uncertitudes of the future. In our introduction, we underlined the fact that one must turn around Benjamin Franklin's saying. Money is time because it means projects.[26] Money permits us to act, to reassure, and to anticipate. It gives us the possibility to secure the future and to act on the future. Money makes us able to wait, to confront an "undetermined future" calmly.[27] This is the basis of money's great power. If we base ourselves on this fact, it is possible to understand what credit is about (in some aspects credit is similar to capital). Credit permits us to have money at our disposal—so we can realize our projects, so we can acquire time. Borrowed time with borrowed money!

Credit "defuturizes" the future. One can possess in the present things that one would not be able to get until much later in life, if ever. In this way we can understand Helga Nowotny's "extension of social present."[28] The future can only be a continuation of the present, but if one wants things before time, one must pay a price. The indebted person—the new type of "money man"—pays for today's happiness with the obligation to pay back his debts tomorrow. This is a situation that can become most worrisome; there are always installments to pay. The future takes on a new form of uncertainty: the possible impossibility of not being able to pay back. The future makes the present uncertain. The indebted person will try to render commonplace the phenomenon of debt, an attitude that makes public indifference only worse.

The economics of credit have provoked considerable turnarounds in certain situations. Irreversibly we have gone from "saving capitalism" to "loan capitalism."[29] In the new logic of economics, "the old ascetic production and accumulation morality based on savings, counting, abstinence, sobriety has been replaced by a hedonistic morality based on credit, spending, pleasure."[30] This type of economics is very precarious. Dahrendorf speaks of "as if" economics where only a dim line separates luxury and bankruptcy, and what is even worse, the "game of chance with the future." The slogan goes: "Have everything right now, and we will see later how we will finance it."

By consequence there is not only a loss of value, more precisely a loss of reality of values that becomes doubtful by the inflationists' use of them, but also the creation of real "no-future" societies. The future leaves no promises, only debts. Dahrendorf is persuaded that "economic reality and human attitude have simultaneously destroyed the time perspective of human life."[31]

On the other hand, the credit institution follows the logic of security: "The feeling of personal security that the possession of money gives is perhaps the most concentrated and pointed form and manifestation of confidence in the socio-political organization and order."[32] This logic has permitted the creation of the welfare state. The basis is again the anticipation of the future mortgaged at the same time, a mortgage that is counterbalanced on a personal level by immediate happiness and pleasure (which is linked to the wish of distinguishing oneself) and on the level of society by stability and social order.[33] Humans are closely linked to the system by credit, the happiness it brings, and the debt they accumulate as a consequence. We cannot avoid, and this is a first conclusion, the difficult question of "happiness." "Money isn't happiness," says the moralist. But that is a myth. Should not one use this odious phrase like Hans Blumenberg does: "To have money is the only determined form of happiness that we can speak about?"[34]

Our discussion suggests another hypothesis: indifference as a result of a life of credit. Money, though it does not leave any person indifferent, is responsible for a more general indifference within modern society. "Money is the mirror for the society, at the same moment it is reflecting into the world its own 'characterlessness' and 'indifference'."[35] Having no other future than that of paying back the present, obtaining everything immediately with one simple signature can be interpreted as a sign of modern hedonism. But the result is, in fact, that everything becomes indifferent and this indifference smothers pleasure.

Notes

[1]Hans Blumenberg, "Geld oder Leben," *Äesthetik und Soziologie um die Jahrhundertwende: Georg Simmel*, ed. Hans Böhringer (Frankfurt am Main: Vittorio Klostermann, 1976), pp. 121–34, quote on page 129.

[2]Isaac Joseph, *Le passant considérable* (Paris: Librairie des Méridiens, 1984), p. 75.

[3]Marc Abelés, "Le mythe contre le désastre," *Le Monde*, November 13, 1987, p. 2.

[4]The 1987 Black Monday is also an example of this relationship.

[5]Max Horkheimer, *Dämmerung* (Frankfurt am Main: Suhrkamp, 1974), p. 235.

[6]See Klaus Heinemann and Peter Ludes, "Zeitbewusstsein und Kontrolle des Zeit," ed. Kurt Hamme and Michael Klein, *Soziologie des Alltags* (Sonderheft 20 of *Kölner Zeitschrift für Soziologie und Sozialpsychologie*, West Deutscher Verlag, 1978), pp. 220–43.

[7]See Vittorio Mathieu, *La filosofia del denaro* (Rome: Armando, 1985). See also Gonzague Pillet, ed., *Les donneurs de temps* (Albeuve: Castella, 1981).

[8]Henri Guitton, *La Monnaie* (Paris: Dalloz, 1974), pp. 17–18.

[9]Georg Simmel, *The Philosophy of Money*, trans. Tom Bottomore and David Frisby (London: Routledge & Kegan Paul, 1978), pp. 510–11, (first published in 1900) (Berlin: Duncker & Humblot, 1977), p. 583.

[10]Claude Javeau, "Vie quotidienne et méthode," *Recherches sociologiques*, 16 (Louvain, 1985), 281–92. See also Claude Javeau, "Huit propositions sur le quotidien," *Sociétés*, 3 (March 1986).

[11]Fernand Braudel, *Civilisation matérielle, économie et capitalisme XV-XVIII siècle*, 2 (Paris: Arthaud, 1985), p. 383.

[12]See Laurent Chemineau, ed. *L'argent invisible. L'ère des flux électroniques* (Paris: Autrement, 1987).

[13]Simmel, *The Philosophy of Money*, p. 198 (German, 190).

[14]Simmel, *The Philosophy of Money* p. 179 (German, 166).

[15]See Ugur Muldur and Nezih Dincbudak, *La monétique* (Paris: Editions de la Découverte, 1987), p. 36.

[16]See Henri Ahlborn, *Cours élémentaire de technique bancaire* (Luxembourg: Editions de la Chambre de Commerce, 1986).

[17]Robert S. Lopez is quoted in Jürgen Rinderspacher, *Gesellschaft ohne Zeit* (Frankfurt: Campus, 1985), p. 38.

[18]See Jacques Le Goff, *La bourse et la vie* (Paris: Hachette, 1986).

[19]Simmel, *The Philosophy of Money*, p. 479 (German, 544).

[20]See Dieter Claessens, "Rationalität revidiert," *Küolner Zeitschrift für Soziologie und Sozialpsychologie*, 17 (1965), 465–76.

[21]Simmel is quoted in Bruno Accarino, "Vertrauen und Versprechen [page 130]. Kredit, öffentlichkeit und Individuelle Entscheidung," in *Georg Simmel und die Moderne: neue interpretationen und materialen*, ed. Heinz-Jürgen Dahme and Otthein Rammstedt (Frankfurt am Main: Suhrkamp, 1984), pp. 116–46.

[22]Simmel, *The Philosophy of Money* p. 177 (German, 163).

[23]Niklas Luhmann, *Vertrauen ein Mechanismus der Reduktion Sozialer Komplexität* (Stuttgart: Ferdinand Enke Verlag, 1968).

[24] J. David Lewis and Andrew Weigert, "Trust as a Social Reality," *Social Forces*, 63, no. 4 (1985), 969.

[25] Claude Javeau, "Les symboles de la banalisation," *Cahiers Internationaux de Sociologie*, 75 (1983), 341–53.

[26] See Mathieu, *La filosofia del denaro.*

[27] Niklas Luhmann, "Wirtschaft als Soziales System," *Soziologische Aufklärung, I*, 1 (Opladen: Westdeutscher Verlag, 1971), pp. 204–31, quote on page 214.

[28] Helga Nowotny, "Time Structuring and Time Measurement: On the Interrelation between Timekeepers and Social Time," in *The Study of Time*, 2, ed. J. T. Fraser (Berlin: Springer-Verlag, 1975), p. 330.

[29] Ralf Dahrendorf, "Ökonomie des Als-ob," in his *Reisen nach innen und aussen Aspekte der Zeit* (Stuttgart: Deutsche Verlags-Anstalt, 1984).

[30] Pierre Bourdieu, *La distinction* (Paris: Editions de Minuit, 1979), p. 356.

[31] Dahrendorf, *Reisen nach innen und aussen*, p. 81.

[32] Simmel, *The Philosophy of Money*, p. 179 (German, 165).

[33] Jean Baudrillard, *A l'ombre des majorités silencieuses* (Paris: Denoël, 1982).

[34] See Blumenberg, "Geld oder Leben," p. 129.

[35] Klaus Lichtblau, "Die Seele und das Geld," *Kultur und Gesellschaft*, ed. Neidhardt (Sonderheft 27 of *Kölner Zeitschrift für Soziologie und Sozialpsychologie*, 1987), pp. 57–74.

Jeux de Masques: Postmodern Tribalism

Michel Maffesoli

Translated from the French by Charles R. Foulkes

By way of hypothesis, I propose as the subject of my reflection the constant coming and going that occurs between the growth of mass society and the development of microgroups aptly called "tribes."

As this century closes, the fundamental tension characterizing sociality is that the masses, or the people, unlike the proletariat or other classes, do not constitute a group that has been formed through a common concept of identity. Consequently, they are without precise focus and are not the subjects of an ongoing and developing history. The metaphor of the tribe, in itself, permits one to become aware of a process of deindividualization, of a saturation of the function inherent in it, and of an accentuation of the *persona*, that is, the role that every person is called to play in the self. It is well understood that because the masses are perpetually seething, the tribes that crystallize within them are not stable: The people comprising the tribes are able to move from one tribe to another.[1] This continual migration calls for a new framework of analysis. One can understand the shift that is going on and the tension it generates by way of the following diagram.

Since French Sociologists Gilbert Durand and Edgar Morin set forth their ideas,[2] this theme can no longer be avoided by intellectuals.

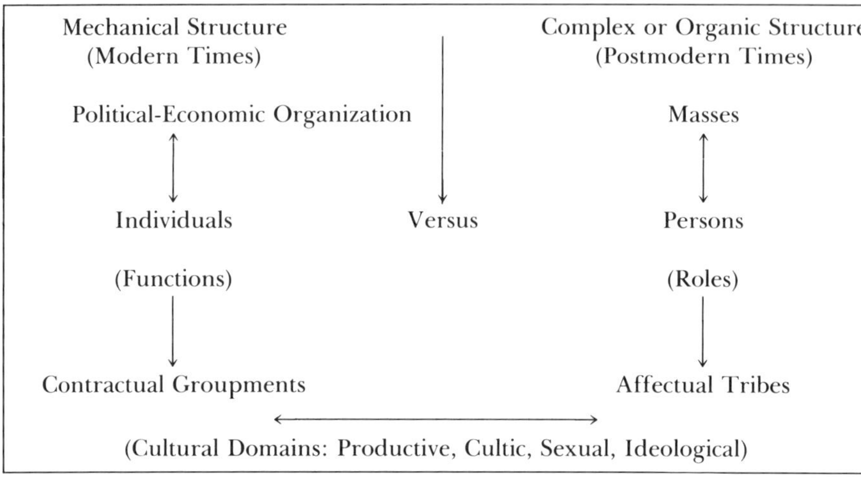

One must recognize that there is a never-ending process underway that leads from the culturalization of nature to the naturalization of culture. This in turn leads to an understanding of the subject in both its natural and social milieu. In light of this, it behooves us to pay attention to the changes going on in our societies. One such change is that the purely rational and progressive Western model, as we know it, has been globalized. It is being saturated and one is witnessing interpenetrations of culture that recall the third term (contradictory) that was just in question. Along with this rapid westernization during the last century, the *orientalization* of the world has begun. Eastern influence is expressed in the fashions of specific life-styles (for example, new ways of dressing), as well as in the emergence of new attitudes regarding both the occupation of space and the body. The Eastern view of the body, in particular, shows itself in the development and the multiplicity of alternative forms of medicine as well as diverse forms of group therapy. Furthermore, ongoing studies point out that, far from being marginal, such practices are making their way into the whole social fabric in various forms. Naturally, this goes hand in hand with the introduction of syncretistic ideologies that, attenuating the classic body–soul dichotomy, surreptitiously elaborate a new "sense of time" (*esprit de temps*), which the sociologist cannot ignore. At just this point, one again picks up this intrusion of the foreign, which is just what Baltrusaitis demonstrated for "Egyptomania," but it seems as if the processes that bring this about are no longer restricted to an elite. They pervade the many little tribes and, by rubbing against one another and intermixing, create the effect of culture.[3]

Now the essential characteristic of the points just made is a new concept of space-time relationship. The emphasis is on the immediate and the affectual, that is, that which is tied to a given place, a place shared with others. A heuristic example comes from Augustin Berque, who states that "it isn't impossible for certain concrete present-day aspects of western culture to incorporate certain aspects of traditional Japanese culture."[4] If one is very familiar with his analysis of this point, one notes that the main point of this regrouping emphasizes the global, the natural, and the relationship to one's surroundings, all of which lead to *behavior of a communal nature.* "The nature-culture and the subject-other relationship are inextricably linked to the perception of space."[5] Strictly speaking, abstracting oneself as little as possible from one's milieu (which in this case must be understood in the broadest sense of the word) tends toward a symbolic perception of one's existence, and existence in the "immediate perceptions and points of reference"[6] will be given preeminence. The relations of the spatial, the global, and the "intuitive-emotional"[7] are completely absorbed in the lost, denied, and discredited tradition of the sociological whole. Such a sense of an organic solidarity, a feeling of all-encompassing roots, which, of course, could never have existed, nonetheless remains the nostalgic foundation that directs or opposes many of our analyses. The notion of *Einfühling* (empathy), which comes to us from German romanticism, best expresses this avenue of research.[8]

As paradoxical as this may seem, the Japanese example could be seen as a specific expression of this whole, this mystical correspondence that comforts society like a *mythos*. In effect, whether it be in business, daily life, or recreation, few things seem to escape it. It seems that the contradictory mix induced by it is not without present-day ramifications at different levels: political, economic, or industrial. This also yields a definite fascination among our contemporaries. Perhaps we should speak, as does Berque, of a "Japanese paradigm,"[9] especially if the term "paradigm," compared with the model, creates a structure both flexible and perfectable. What is for certain is that this paradigm encompasses nicely the mass-tribe dialectic, which is my prime subject here. This dialectic is one of ceaseless and somewhat indefinite movement; its "form" lacks both center and periphery; and it comprises elements that, according to accepted encounters and experiences, adjust into changeable forms based on preestablished archetypes. This swarming, this seething of culture, results in an unsettling of our individualist and individualizing modes of thought.

But is this a new concept, after all? Other civilizations are founded on the ritual interplay of disindividualized persona based on roles collectively lived out that in no way failed to produce stable and relevant social structures (*architectoniques*). Let us not forget that the affectual confusion of the Dionysian myth has produced important civilizing effects; it is possible that our megalopolises serve as a framework for their renaissance.

THE TRIBAL AMBIANCE

I have no intention of directly tackling this problem of individualism, but I will address it *a contrario*. It is essential to point out, describe, and analyze the social configurations that seem to go beyond individualism, and to understand the undefined mass, the population without identity, or the tribalism that, though nebulous, contains small local units. It is certainly a question of metaphors, which tend to accentuate the confusing aspect of the sociality (once again the emblematic figure of Dionysus comes to mind). Hypothetically, I propose to act as if the categories that have served us for more than two centuries in analyzing society were completely saturated. Reality goes beyond fiction, so the saying goes. Let us try, therefore, to keep abreast of the former. Perhaps it should be demonstrated how certain novelists treated the issue, conveying the idea that the individual no longer has the substance once credited to him by philosophers since the Enlightenment.

Samuel Beckett's plays lead the way by destroying the illusion of an individual who is master of both himself and his life. In a paroxysmal and sometimes premonitory manner, he reveals the contingent, ephemeral aspects of all individualism, giving equal emphasis to the factitiousness of the process of individuation, including the fact that it leads to a prison. Individualism is an outdated bunker, one that deserves to be abandoned. This is the picture Beckett gives us, a view that does not lack constructive originality in today's world of ready-made thought. This view, which surely emanates from many such thurifers, is yet perfectly synchronized with the wisdom of antiquity making of each individual a simple *punctum* of an uninterrupted chain, even attributing to him a multiplicity of facets, each constituting a microcosm, a crystallization, and an expression of the general macrocosm. One recognizes here the persona, the mask that can be mutable and that, above all, integrates itself into a variety of scenes and situations whose relevance exists only because they are played out by many.

The multiplicity of the self and the communal ambiance that it instills will serve as the backdrop for our inquiry. I propose to call this the "esthetic paradigm," in the sense of communally experiencing or feeling. In effect, since individualistic logic rests on a separate and self-enclosed identity, the person (persona) is worthless except in relation to others. Sociologically examining some modern authors such as William Faulkner and Thomas Mann, Gilbert Durand addresses this idea of a "force of impersonality," which allows one to exist only in "the spirit of others."[10] This point of view forces us to go beyond the classical dichotomy between subject and object, which serves as the basis of all bourgeois philosophy. The emphasis is on that which unites rather than on that which divides.

The schema that I am contractually constructing is not associated with other rational individuals but is rather a myth to which I adhere. Heroes, saints, and emblematic figures can exist, but only as ideal types, as empty forms or molds, that allow everyone to recognize themselves in an ideal self and to commune with one another. Dionysus, Don Giovanni, a Christian saint, or a Greek hero: The list of mythological figures is endless. All are social types that allow for a common "aesthetic" and function as receptacles for an expression of "we-ness." The multiplicity of this or that type inevitably favors the emergence of a strong collective sentiment.

At certain moments the "divine" society takes shape around a collective emotion that recognizes itself within a certain type. An artistic or political theoretic genius may have put forth an idea whose essence indicated the path to be followed. The ones that put forth a goal to be attained remained abstract, inaccessible entities. The proletariat or the bourgeois could be seen as "historic subjects" that had a goal to attain. In contrast, the mythological type has a simple function of aggregation; it is solely a "holding tank." It only expresses, for a fixed span of time, the collective genius. Here lies the difference between an abstractive/rational period and an empathetic time. The former relates to the principle of individuation or separation, while the latter is dominated by indifferentiation, the "losing" of self into a collective subject. This indifferentiation is what I refer to as *neotribalism*.

Numerous examples in our daily lives serve to illustrate the emotional atmosphere brought on by the tribal development; to no one's surprise, it is part of the urban landscape. The diverse sights of "punk," "kiki," or "paninari," which nicely express the uniformity and the conformity of these groups, offer specific instances of the ongoing spectacle generated by contemporary megalopolises. One can link the tendency toward the visible orientalization of life in Western cities with Augustin Berque's analyses of the relationships of "instinctive attraction" between "self" and "other" in Japan. The weakness of this distinction between self and other—subject and object—is what leads to reflection. The concept of the extendability of the self ("a relative and extendable ego") can be a most important methodological tool for the understanding of today's world.[11] One hardly needs to mention the fascination that Japan evokes today or remind oneself of its economic and technological performance to emphasize the fact that if the *distinction* between self and others can be a notion that applies to contemporary life in general, it is, on the other hand, totally inadequate to describe the diverse forms of social aggregation seen today. Such groups have indefinite parameters—sex, appearance, fashion, even ideology—all of which are more and more often qualified in terms such as "trans" or "meta" that transcend the logic of the unitary or the binary. In summary, by assigning these terms their strongest meaning, one can say that one is participating in the replacement of a rationalized *society* by a basically empathetic *sociality*.

This process expresses itself in a succession of moods, sentiments, and emotions. For example, the German enlightenment notion of *Stimmung* ("atmosphere") serves a greater and greater function, first in describing the reigning relationships within social microgroups and second in specifying the manner in which these groups situate themselves in their spatial environment in terms of ecology, habitat, and neighborhood. Similarly, the constant use of the English term "feeling" in the field of interpersonal relations merits attention. It will serve as a criterion for measuring the quality of exchanges and for deciding whether they ought to be pursued, as well as the degree to which they should be intensified.

In effect, it seems necessary to make a change in our modes of evaluating social regroupments. In this respect, we can use Murray Webster's sociohistorical analysis of the *Gemeinde* ("emotional community") to our benefit. He points out that we are dealing with a "category," that is, something that never existed in and of itself but may serve as an indicator in present situations. The emotional community is characterized primarily by an ephemeral quality, a "fluid composition," a "lack of an organization," and a rather "fixed routine" or *Veralltäglichung*.

BEYOND THE NOTION OF IDENTITY

It goes without saying that the metaphor of tribalism, or even of "emotional community," goes beyond the categories we are accustomed to using in our analyses, especially in dealing with the logic of identity. Indeed, even though the notion of identity has served as the pivot for the economic, political, and social order of the last two centuries, and even if it continues to function, its steamroller-compressor effect no longer has the same efficacy. Hence, the best way to grasp fully the shared feelings and experiences at work in various social situations and behaviors is probably to attack the problem from a different angle, that of the esthetic. (I am using "esthetic" in an etymological sense as the communal ability to feel or experience.) In spite of his rationalism, Theodor Adorno remarked that the esthetic could allow for the "defense of the non-identic which in reality suppresses the constraint of identity."[12] One can hardly better emphasize the outbreak and exuberance of neotribalism. In its various forms, neotribalism refuses to be identified with specific political endeavors, does not conform to any single definite structure, and has as its sole *raison d'être* the preoccupation with the collectively lived present. Clear evidence of this abounds in the research and monographs done on groups of young people, on cliques, on small industrial enterprises, and on fashion circles.

The various lamentations of politicians, the clergy, and journalists over the growing disindividuation are an indication of "suprasingular" or "supraindividual" realities. All normative evaluation aside, one must be able to extract the various consequences. To start with, regarding psychological experiences in the 1970s, Paul Watzlawick spoke of "the ardent and unshakable desire to be in accord with the group." Actually, it is not even a question of desire anymore, but rather of the ambiance in which one chooses to bathe. What was once experimental in the California groups has become common reality in everyday life. The desire still referred to a subject of which it was the carrier; this is no longer the case. The concern with conformity is a consequence of the process of massification, a process within which regroupments occur in a somewhat random, aleatoric manner. Above I referred to the "materialism" of the communal body, of which the coming and going of the mass-tribe is an example. One can imagine that instead of encountering a subject-actor, one is confronted with an entire cast of objects, the whole object-mass encompassing small groups that subdivide *ad infinitum*, like a Misha doll.

In elaborating his ethic of sympathy, Max Scheler sets out to show that it is neither essentially nor exclusively social but rather of an all-encompassing or even matrix-like form. And such is the kind of hypothesis that I, in turn, will formulate. According to the pendulum of history, this form would be newly arrived. It would favor, as natural outgrowths, emotional functioning as well as mechanisms of identification and participation. What he calls the "theory of identification of sympathy" allows us to understand the instances of fusion, the moments of rapture that can be punctual yet can equally characterize the climate and the spirit of an age.[13] This theory of identification, this ecstatic sortie from the self, is in perfect congruence with the development of the image, be it the image of a spectacle (strictly seen as a political parade), the crowd at a sports match, a group of tourists, or just a mob of idlers. In all such scenarios, one can participate in a "going-beyond" of the *principium individuationis*, which has been the "golden mean" of all social schematization and theorizing until now.

Is it necessary to establish, as Max Scheler suggests, a gradation between affective "fusion," "reproduction," and "participation"? In my opinion, a better and more heuristic approach would be to establish a nebulous *affectuatio* of orgiastic or, as I discussed above, of Dionysian tendency. Be they orgiastic explosions or cults of possession, fusionistic occurrences have always existed. But sometimes they take on an endemic quality and become preeminent in the collective conscience. Whatever the subject may be, we are on the same wavelength. Maurice Halbwachs speaks in this regard of "collective interferences."[14]

This nebulous *affectuatio* allows us to understand the specific form that the sociality assumes these days: the coming and going of mass-tribes. In effect, in contrast to the prevailing tendency in the 1970s (its strengths being the Californian counterculture and the European student communes), we are dealing less with aggregating into a group, a family, or a community, and more with zipping from one group to another. This is what can create the impression of an atomization and also wrongly give rise to talk of narcissism. In fact, contrary to the stability induced by classic tribalism neotribalism is characterized by fluidity, by punctuated gathering and scattering. This fluidity best describes the spectacle that goes on in the streets of modern megalopolises. The jogger, the punk, the retro-look, the Bon-Chic-Bon-Genre, the street entertainers, the rappers all invite us into a world of incessant movement. Through successive layering emerges the esthetic ambiance in question. It is in such an ambiance that these "instantaneous condensations" (Hocquenghem-Scherer), fragile, yet still the objects of a strong emotional investment, operate. This sequential aspect enables one to speak of going beyond the principle of individuation. Let us illustrate. In a nice description of American highways and their traffic patterns, Jean Baudrillard notes the peculiar ritual of the "regularity of its flow which eradicates individual destinations." For him, the "only veritable society, the only *chaleur* here, is that of propulsion, a collective propulsion."[15] This image can help us think. In a quasi-animal way, we sense a force that transcends the individual trajectories, or, rather, one that inscribes these in a vast ballet, in which the participants, stochastic as they may be, form nothing less than a constellation in which the various components adjust without will or conscience being an issue. Such is the arabesque of the sociality.

The characteristics of the social are that the individual can have a *function* in the society and also function within a party, an association, or a stable group. The characteristics of the sociality are that persons play *roles* within their professional activities, nurturing the various tribes in which they participate. Their stage costumes changeable, they go out every day, according to their fancy, whether sexual, cultural, social, or religious, to take their place in the diverse games of the *theatrum mundi*.

One can never overemphasize the contrast between the social and the tragic superficiality of the sociality. I have already demonstrated, with regard to daily life, how the profound could hide itself in the surface of things—hence the importance of appearance. It is not a question of going into such an issue in and of itself but rather of briefly noting it as a guideline of aggregation. In the sense indicated above, the esthetic is a means of testing, of feeling in common. It is also a means of recognition.

Parva esthetica? At any rate, the variety of modes of dress and multicolored hair, as well as other manifestations of punk, function as cement. The theatrical founds and comforts the community. The cult of the body and games of show are meaningful only because they are part of a vast scene in which everyone is at once both actor and spectator. To paraphrase Simmel and his sociology of the senses, it is a matter of a scene that is "common to all." Less emphasis is put on that which distinguishes than on the universality of the effects.[16]

The nature of the spectacle is to accentuate, directly or euphemistically, the sensitive-tactile dimension of the social. Togetherness facilitates touching. "Most popular pleasures are crowd or group pleasures," according to A. S. C. Ehrenberg, and one cannot understand this strange compulsion to assembly if one does not have this ongoing anthropology in mind. I come back to the dichotomy developed by Wilhelm Worringer between abstraction and *Einfühlung*: There are moments, abstract, theoretical, purely rational, when the culture, in its broadest sense, is made up of participation and "tactility." The reappearance of the image of the sensitive in our society certainly appeals to a logic of touching.

THE GAME OF THE COLLECTIVE IMAGE

In conclusion one must establish a parallel between the collective sensitivity and the importance that the imagined reacquires in social life. One almost needs to establish a sociological "law" on the matter: Whenever mistrust regarding the image tends to predominate (iconoclasm or rationalistic monovalence), theoretic representations and forms of social organization that have the "remote" as their common denominator develop, and one takes part in the domination of politics and historical linearization, both of which are essentially prospectives. In contrast, whenever the image, under the influence of these various modulations, returns to the forefront, localism becomes the unavoidable reality.

One historical example can serve as a springboard for our analysis. At the time of the founding of Christian civilization, *iconoclasm* was the ideological barrier under which the tenets of centralism formed, whereas *iconodualism* is the way of those who give privileged rank to expression of local interests. Of course, a theoretical rationalization, theological in nature, is given to this conflict, but the important thing is to know what form the organization of society will adopt. And Peter Brown, who analyzes this conflict, even refers to "*Jacobinist iconoclasm.*" All the means were good for extirpating local cults, simply because they inhibited the activities of the central government. These local cults formed around a sainted person and a specific icon; hence, "one or the other received its consecration from below." From this, an entire complex system was erected of interrelations between the various "topoi" of a veritable parallel society that had escaped the centralized organization being established.[17] From this process we can see that the role of the icon is one of legitimizing the counterauthority of the sainted person and of serving as the crystallization of the feelings of the local groups.

In the solitude inherent in all urban areas, the icon is a familiar image—a sort of group "uniform" (brandy, punk, retro-look, and so on)—that acts as a point of reference and becomes a part of daily life. It is the center of a complex symbolic and concrete order in which everyone has a role to play within the framework of a global stage. It also allows for the recognition of oneself by oneself and by others, and finally, of others by oneself. This is where the thrust of the empathetic image lies. The empathetic image resurges regularly in order to palliate the mortifying effects of uniformation and of the environment it induces.

Naturally, it is appropriate to appreciate that there may be contemporary modulations of what we have called icons. These are diverse, and every one of them would mandate specific icons. Here I have confined myself to revealing the workings or, once again, the "form." But these should allow the uncovering of the "imagined" function of a great many local symbols. These in turn can be noteworthy items of whatever type: the animals with which a group identifies itself, a specific place, the products of the soil, or, once again, specific ways of dressing.

One can add that the power of the emblematic image has accrued with technological development. Originally suspected of being the global carrier of an ideological and alienating message, television is no longer the carrier of a single and discrete message meaningful to all. Instead, television gears itself to target audiences, tribes that arise and identify with archetypal images representing products, goods, services, or lifestyles typical of the group or tribe. In effect, even if what we here suggest is merely a tendency, one must be aware that it applies more and more to particular groups identifiable by age, region, town, even neighborhood. Other examples, including "cabled" apartment houses, can only reinforce this process. Even if the image is further away, heavier, or totally abstract, it is nonetheless manifest in proximity to the viewer. The merit or value of the image is not the issue; it will play the role of the familiar icon regardless.

An apartment building or a neighborhood presents itself as a spectacle. In the framework of the megalopolis the televisual image becomes a part of the tactile, emotional, and affectual rapport. In this sense, it confronts the tribe in such a manner that it creates for itself a secure place, One sees that the theoretic stakes are high and that, if one pays attention, it is "from the ground up" that these new manifestations of the "*être-ensemble*" move up.

What is for certain is that all of this relates to space. The various examples given indicate a territorial connotation. On the basis of linguistic research, Berque makes a distinction between "egocentric" and "lococentric" languages.[18] It is certainly possible to extrapolate his analysis and to realize that some courses are predominantly "egocentric" while others are "lococentric." The former privileges the individual and his concerted actions, whereas the latter places the emphasis on the environment, be it natural or social. One can equally imagine that within the same culture one finds differing sequences, the accent sometimes on that which individualizes and at other times on that which deindividualizes the collective aspect. At any rate, this is my hypothesis regarding our times. In this sense, the valorization of the space, of the body or group feeling, through the bias of an image would simply be the cause and the effect of the going-beyond of the individual in a greater whole. A society based on such a dynamic risks seeing its essential values reversed; this is, perhaps, the modern failing of all experiences and social encounters based on the sociality.

Notes

[1]The individual is closed, self-contained; the person is above all a mask. The nature of the person is that he or she is polysemantic, polyphonic, as I suggest later on.

[2]See Ron Levy, "Critical System Thinking: Edgar Morin and the French School of Thought" *System Practice*, vol. 4, no. 2, 1991, pp. 87–99.

[3]Jurgis Baltrusaitis, *La quête d'Isis* (Paris: Perrin, 1967).

[4]Augustin Berque, *Vivre l'espace au Japon* (Paris: PUF, 1982), pp. 34, 31–39.

[5]Berque, *Vivre l'espace au Japon*, p. 85.

[6]Berque, *Vivre l'espace au Japon*, p. 37.

[7]Berque, *Vivre l'espace au Japon*, p. 32.

[8]Regarding "Einfühlung," see Michel Maffesoli, *La congvête du présent* (Paris: PUF, 1979).

[9]Maffesoli, *La conguête du présent*, p. 201.

[10]Gilbert Durand, "Le retour des immortels," *Le Temps de la Réflexion* (Paris: Gallimard, 1982), pp. 207, 219.

[11]Berque, *Vivre l'espace au Japon*, p. 54.

[12]Theodor W. Adorno, *Aesthetic Theory* (New York: Routledge Chapman E Hall 1986), p. 13.

[13]Max Scheler, *La nature de la sympathie* (Paris: Payot, 1928), pp. 13, 83, 149–52.

[14]Maurice Halbwachs, *La mémoire collective* (Paris: PUF, 1968), p. 28.

[15]Jean Baudrillard, trans. Chris Turner, *America*, (New York: Routledge, Chapman & Hall, 1987) (Paris: Grosset, 1986).

[16]Guy Hocquenghem and René Scherer, *L'âme atomique* (Paris: Albin Michel, 1986), p. 25. See also Wilhelm Worringer, *Abstraction and Empathy* (trans. Michael Bullock, Independence, Mo.: International University Press, 1963).

[17]Peter Brown, *Society and the Holy in Late Antiquity* (Berkeley: University of California Press, 1989). pp. 224–26, 218.

[18]Berque, *Vivre l'espace au Japon*, p. 47.

Toward Digital Man?

Victor Scardigli

Translated from French by Caryn Bensmida

Today's technology builds tomorrow's humanity. It is the key to most forecasts about the impact of scientific and technical innovations on work and the organization of worldwide exchanges; on urbanism and social life of the city; on value systems and methods of thinking; on cultures; and, perhaps, even on the human species.

What is the value of a futuristic scenario of a digital and immaterial society? In evaluating this question, caution is required: Imagination exceeds actual changes. But I will insist on the benefit of studying, among other matters, this fantasized future that guides the action of all partners in the spread of the current technology, which combines information science, audiovisual media, and telecommunications.

FROM NEANDERTHAL MAN TO DIGITAL MAN

A galaxy of telecommunications, the computerization of society, the mastery of the living: Each decade brings its discoveries, stupendous and terrifying, and humanity immediately begins again to hope for, or to fear, a social revolution produced by the technical revolution.

Among the prophecies that accompany scientific and technical advancements, the most well known are of those who promote this progress. Scholars, industrialists, politicians, and journalists often develop a deliberately optimistic vision of the future announcing the technologically ensured happiness that awaits man at the dawn of the third millennium.

An often considerable gap separates these expectations from their realization, as will be shown in recalling the prophecies of the 1960s and 1970s. Other scenarios, much less technical or happy, are at least as plausible: They have been sketched in *La société digitale* (*The Digital Society*).[1] The experts' and the mass media's predictions are allied most often with a similar dominant scenario of linear, coherent, even problem-free evolution toward an "information society," and then, a "digital" humanity.

According to Porat, the increased percentage of "information workers" marks the passage to the postindustrial society.[2] In primitive and, later, industrial societies, the value of products consisted of raw materials and manual labor; in the new order of society, economic and then social value consist chiefly in the progress of knowledge incorporated in the new objects and services presented to consumers. In this sense, mass media, telematic services, and other consumer information mark the "overall change from hardware to software forms," to draw out the formula of McLuhan. Following Porat, Americans and Japanese (such as Daniel Bell[3] and Yoneiji Masuda[4]) consider the information society to be the postindustrial society, whereas the European sociologists are more reserved regarding these two concepts.

I prefer to speak of *techno-pole*, that is, an order of society founded on an indissolubly commercial definition (*pole*, in Greek) and technical definition (*techne* designates high professional competence, knowledge at the service of action) of human needs and their satisfaction (Scardigli, 1983[5] and 1985[6]). Information is therefore only one facet of the techno-pole society of the twentieth century. But this facet is becoming more and more important, along with the generalization of human activities that use a numerical processing for sound or light signal, massed data, and, progressively, the whole of physical and social reality.

Although electronic calculators have been perfected since the 1940s, in the mid 1970s they experienced a sudden expansion of performance and application, notably due to the invention of the microprocessor. Meanwhile, the costs have been dropping. Microchips are found everywhere, from automobiles and traffic control to household appliances and credit cards. New types of words—*bureautique* (computerized office work), robotics, even *educatique* (computer-assisted instruction), *productique* (computerized production), and *monétique* (computerized monetary transactions)—mark the computerization of society predicted by Simon Nora and Alain Minc.[7]

Alvin Toffler contends that we are in the midst of a technological revolution as important as that of the printing press and the steam engine,[8] thereby agreeing with the diagnosis of Radovan Richta,[9] the socialist prophet of the scientific and technical revolution. And for many people, information science, associated with the progress of telecommunications and audiovisual media, changes cognitive psychology and philosophy as much as sociology.

Until the invention of the computer in the 1940s, the representation of the world that technology adopted was analogical, as shown by Philippe Breton.[10] First human tools, then automatons, and finally cybernetics were applied to reproduce human gestures, the appearance of behavior of living beings; information science, on the other hand, offers a purely abstract usage. In speaking of the digital society more broadly, Pierre-Alain Mercier[11] suggest that the relationship of people to the world is progressively governed by digitalization, that is to say, the numerical processing of signals. A sound or light signal becomes a basal system of numbers that can be stored, transmitted, and reproduced indefinitely, but also invented or transformed. Henceforth, voice and sight, time and space, thought and action are digitalized. Even genetic inheritance ends up being reduced to a code: Five amino acids, the same for all living beings, come to defy the followers of the transcendence of the human species and the qualitative specificity of each person. Humanity is summed up by the order in which these five pieces of data are sequenced.

This concept of digital society thus opens a combination of questions and reflecting.

A reflection of a philosophical nature on the relationship of people to the world. What is reality, when an engineer can produce a synthesized picture or an artificial voice that has all the semblance of reality? Jean Baudrillard,[12] Marc Guillaume,[13] Jean-François Lyotard[14] think that we are constructing a society of widespread spectacles, artifice, and false appearance. Already the technological evolution aboard aircraft imposes a digitalized representation of the world surrounding the pilot. Instead of looking outside the cockpit and maneuvering with his arms and legs, now the pilot's function is to keep an eye on the parameters provided by the computer on board.[15]

A new set of sociological questions. Robotics, bureautiques (computerized office work), and the progress of expert systems toward artificial intelligence seize the lessons of experience from the highly qualified worker and rob them of their competence. These systems upset the division of labor and could lead to a worldwide employment crisis.

A cultural anthropological reflection. The mentality, value system, and life-style of a society that brings into general use artificial substitutes for working, moving about, repairing the body, and, soon, for thinking and living will be affected. Logical mathematical reasoning seems to be in opposition with the intuitive, total, and qualitative conduct of "the man in the street." The irrational and the emotional, as well as oral recollection, have no place in the rationalization of data banks and telecommunications. The *domotique* (computerization of the home) introduces this technical trade and digital mentality into private life, in the very heart of one's dwelling.

A reflection about physical anthropology and biology. The anthropologist André Leroi-Gourhan[16] did not hesitate to outline a parallel between the first human tool and today's artificial memory. The gesture of the primate grasping a stake or a pebble tool in his hand aroused human intelligence; at present, data banks, dismissing memory from the brain, free our neurons for new functions.

All of these questions are fundamental, and it is important that the social sciences seek solutions.

Before examining the meaning of the first applications of new technology, the fact that they are a magnificent projective test (psychological device) for researchers should be noted. After all, researchers participate, as does everyone, in the same collective imaginary vision of progress: They would not know how to break away entirely from this modern *Weltanschauung* in which the future will necessarily be dictated by sciences and technology.

DISILLUSIONS, REPRODUCTIONS

More than ten years have passed since the experts proclaimed a flash computerization of the planet, and yet this radical upheaval has not occurred. Must the conclusion therefore be drawn that we are not moving toward a digital society? In fact, we are in the second phase of the spread of this technology, that of disillusionment and social reproduction. But close analysis of numerous social experiments of products and commercial services identifies three very distinct phases in the process of social diffusion.

The first phase is that of prophecies and great maneuvers. In the 1970s, the information and communication revolution proclaimed both miracles and catastrophes as follows. Microcomputerization would bring individual autonomy and decentralization. Telematics and local networks would reconstruct local democracy, socialization, fellowship, and lost communication. Video would rouse personal creativity. Neighborhood clubs would allow all to share the technical culture and the end of social inequality. Otherwise, the screens and the robots would reinforce the subjugation of workers and consumers, provoking an economic crisis and social void, and would reinforce the dependence of the Third World.

These predictions were not based on any fact, as there was nothing yet to observe. Nevertheless, predictions can be found everywhere: in the experts' reports, in the rationale for public decisions, as in the arguments of industrialists or the initial publicity. An even more surprising fact is that multiple resemblances appear among the representations specific to each technological area; biology gives rise to hopes and fears in a manner very similar to the technology of information (Tables 1 and 2). A truly collective imaginary vision reappears regarding each advancement of science and techniques.

In a parallel way, the principal economic and political players have a presentment that these innovations will constitute a new stake (or maneuver in order to form them into stakes), and they begin to deploy a strategy for medium or long term, in which phantasms come to argue the decisions.

With respect to applications in private or neighborhood living in this first stage of prophecies, the possibilities do not correspond with any social necessity or any already established daily practice. At the same time, in France, as one example, industrialists and public authorities are convinced that creating a vast domestic market is important to support industry (reasoning developed during the programming of the electronic telephone directory and the Telematic Plan) and to serve as a showcase for exportation (justification for the Biarritz experiment of fiber optics and the *visiophone*—otherwise known as *vidéophone*). Only this computerization society will spare France from underdevelopment and technological and cultural colonization by the United States and Japan. In order to create this large-scale public market, marketing the informal economy is necessary to induce consumers to pay for that which was free: information, communication, leisure, and social relations.

In the Nora and Minc report, sociology was invited to join in the study of these issues to help engineers and planners find uses and to establish connections between senseless products and newly perceived needs stemming from life-style transformations.

In the 1970s, the first video materials, and, later, the first local cable television networks gave rise to genuine social experiments: They served as support in the research of alternative life-styles in a few French villages

TABLE 1 Information and Communication Technology: The Imaginary Vision of Its Prospects and Stakes

The Seven Stakes of Progress	*Hopes and Expected Miracles*	*Fears and Dreaded Evils*
1. The power over constraints: liberty or slavery	End of painful, menial, repetitive labor; extinction of the proletariat (factories without workers)	Telematic dictatorship (Big Brother); "bureaucracy aided by computer"; military imperialism
	End of irksome household tasks (electronic house, intelligent robots)	Man slave of the robot (watching tasks in computerized plants; teachers becoming servers for teaching machines)
	Spatial autonomy (the portable microcomputer); abolition of space and time (telecommunications); omnipresence	
	Gaining time, to the benefit of personal creativity	
	Choice of personal destiny	
	Freedom of expression by minorities (local radio and television)	
2. Knowledge: intelligence or lack of culture	Encyclopedic access to knowledge (large computers, data banks)	Decline of intellectual level (computer babies; end of written work)
	Technical culture	End of artistic creation (artificial images)
	Artificial intelligence serving people, for mastery of the rising level of complexity	
	More intelligent children from exposure to computers	
3. Death: immortality or insecurity	Eternal collective memory (video libraries and so on)	Failure of computers (nuclear apocalypse)
		Error of range finders (catastrophes)
		Violating privacy with audiovisual technology in homes

4. Social justice: equality or injustice	Progress accessible to all with neither monetary nor cultural obstacles: decreasing costs of microcomputerization, public financing of telematics; apprenticeships at school; computers using natural languages	Economic and cultural inequality of access (information science, minitel)
5. Social ties: fellowship or social void	Human communication reestablished (networks) Neighborhood fellowship rekindled (video and local telematics, community television) Decentralization of the territory (country living facilitated by telematics)	Telematics emptying the streets; screens eliminating need for face-to-face encounters Information science repressing the irrational Workers isolated in front of their screens
6. Abundance and enjoyment: welfare state or crisis	Satisfaction of all desires: decrease in costs (electronics) responds to rise in demand and desire Nonpolluting and energy-conserving industry Cultural industries come out of crisis, new area of growth	Unemployment worsening Decline of Europe Crisis in our value system
7. Planetary solidarity: development or devastation of the Third World	Technological shortcut to catch up with the West	Technological trench Widening of international division of labor Computerization destroying cultural diversity

Source: Victor Scardigli, *Imaginaire Progrès* (Paris: CNRS-MCD Report, 1987).

or neighborhoods and especially in Quebec, Canada. Small groups, often prompted by a sort of utopia influenced by the events and attitudes of 1968, attempted to invent a better, more friendly society, opposing inequality and solely productivity-oriented behavior.

Lacking a means of comparison in this first phase, the assertion was made that computerization would be the basis of primary instruction

TABLE 2 The Imaginary Vision of Technology of the Living

The Seven Stakes of Progress	*Hopes and Expected Miracles*	*Fears and Dreaded Evils*
1. The Power over constraints: liberty or slavery	Autonomy of the ill or disabled (prosthesis, ambulatory care) Social and sexual liberation ("the pill") Choosing offspring (sperm banks, genetic engineering) Choosing the time for pregnancy	Life dependency on artificial implants and limbs Dictatorship of "pill" prescribers or of proponents of cesareans Social manipulation through drugs Genetic reprogramming of certain populations
2. Knowledge: intelligence or lack of culture	Gifted embryos The superintelligent human species	
3. Death: immortality or insecurity	A longer life span Complete medicine (cure-all) Expanded program of vaccinations	Threats to human species, devastating virus, genetic manipulation (degenerated mutations) Depradation of the body (chemotherapy)
4. Social justice: equality or injustice	Equal access to new treatments/socialized medicine "Right to bear children" Equality of women (society of test-tube babies)	Inequality entered into genetic pool Human clones
5. Social ties: fellowship or social void		Isolation of patients (AIDS sanitariums); families and friends stripped of primary role in helping patients cope with agony Sexual immorality; destabilization or end of the family (*in vitro* fertilization and surrogate mothers)

6. Abundance and enjoyment: welfare state or crisis	Active old age	
7. Planetary solidarity: development or devastation of the Third World	End of world hunger (green revolution, biotechnology for food provision)	Pillage of traditional medical methods Biotechnology debunking traditional forms of agriculture Demographic explosion ruining poverty-stricken countries

Source: Victor Scardigli, *Imaginaire Progrès* (Paris: CNRS-MCD Report, 1987).

and the use of computers would extend to schools for professional training and remote instruction. At the same time, users would become producers of information, instead of passive consumers—that is, producers of images and ideas as a result of computer-assisted creation. The microcomputer would bring about the blossoming of children who were having difficulty, the "springing forth of the mind";[17] it would allow the residents of disadvantaged neighborhoods and poor countries to overcome their handicaps, such as in the endeavors of the Worldwide Center of Information Science started by J. J. Servan-Schreiber in 1981 in Paris. The *télé-alarme* would change the lives of the elderly. (The *télé-alarme* is a telephone service that alerts the emergency alarm exchange, which calls back instantly when a button is pressed. If there is no reply, help is sent to the subscriber's home.) City life would be transformed by setting up technical networks of local information; true social, interactive communication would finally be reestablished through interaction, the key word of the new technologies. It is, for example, the launching of the local interactive and convivial telematic network ASPASIE at Marne-la-Vallée, near Paris[18].

The second phase began approximately in 1976 for video; approximately in 1983 for microcomputers, television, and local radio; and approximately in 1986 for telematics. An initial evaluation of the spread of new technology outside of business is clearly negative in the area of information science. The public frowns on the use of new technology for other than paraprofessional use (word processing), causing manufacturers to return to professional markets, both small and large firms. Schools have not been any more successful in converting the French to the computer culture. Public authorities delay the heavy financing of telecommunications (in the case of satellites) or cancel projects (for example, the ambitious Cable Plan that was to cover France entirely with a network of fiber optics).

In short, nothing is stirring in French urban life. Telecommunications have not given rise to a massive return to the country; neighborhood fellowship has scarcely reappeared. The so-called social experiments are henceforth urged by public authorities, with the purpose of serving as pretests of telematic and audiovisual products (the community of Vélizy tested the minitel). Ventures such as local Italian radio, so highly boasted, and the diverse community networks in Québec are either commercially profitable in France or they disappear. The only notable change remains technical: There simply occurs a progressive establishment of more and more complex regulations (in urban traffic, domotique (computerized home life) and, soon, "intelligent" buildings).

In private life, there is merely an increase in consumerism; after having purchased small calculators, games, and electronic toys, the public rushes to buy videos, compact disks, and other audiovisual equipment. It spends more time every day before numerous television channels and uses the minitel massively for its multiple free possibilities (directory) or paying options: commercial, financial, information services, and the electronic "mail box"—permitting users to exchange messages.[19]

If new technologies are used by the public, it is only to reinforce the tendencies that already dominate the current life-style. The consumption of audiovisual and leisure products and services is growing, but users do not demonstrate personal creativity with the microcomputer or video equipment; home telematics accentuates the centering of daily living on the individual and the home; neighborhood life and the frequenting of collective urban equipment is declining. And worse: In the appropriation of new goods and services by individuals, there is inequality of access and use related to money, technical culture, and the sexual distribution of tasks. Similarly, in certain countries of the Third World, the setting up of computer and audiovisual programs results in an increased dependency on the United States after five to ten years.

In summary, far from changing society, the new tools seem most often to serve as reinforcements of preexisting uses; they renew inequalities in favor of the already dominant social or political actors. This second phase is that of gadgets and disillusions: The technical revolution seems only to open up an immense social propagation.

THE IMAGINARY IN ACTION

In the second phase of the social development of new technology, we seem to be turning our backs on the digital society, totally imbued with an economic and scientific rationality. Social players forget their own prophecies; the weights of the past reclaim their rights; the technological wave falls back into the banality of the everyday ordinary. A third phase is developing, however. It is what previous inventions teach us that marks our era.

Uses for the telephone were hardly apparent, except perhaps as a leisure gadget: the "theater-phone," by which some privileged persons were directly linked to a playhouse. On the contrary, radio waves were to be used in the interactive transmission of military or economic information between two correspondents. We foresaw neither its interactive, conversational use nor the telematic services that are being installed today. Another example: Trains could be used in the transportation of coal in the mines, but there was the concern that passengers would suffocate in the tunnels or beyond from a speed of 30 kilometers per hour. The automobile was a curiosity reserved for the privileged: Its manufacture required months of work for each unique model, and its technical complexity required the constant presence of a chauffeur-mechanic.

At present, new technologies take less time to put into general use. Certainly, it takes time to transform the world into numerical data: the economic, medical, meteorological, and other data banks are very costly to establish and to update. We have not yet managed to produce images for all of the networks and television satellites in place. But the cost of equipment is dropping. The French jump on the possibilities offered by the minitel: commercial services or electronic "mail box." In Scandinavia or in Italy, TV viewers devote two hours daily to their sets, in France or Japan three hours, and in the United States four and a half hours (Table 3).

And so far we have not spoken of the world of work, where the technical revolution is established much more quickly under the constraints of international economic competition. The problem remains entirely to be seen. Whether in private life or in the community, workers will behave as they have already learned to behave confronted with bureautique (computerized office work) or computer-assisted production. But we note that the confines are lessening between work and leisure and that word processing and, less often, consulting of data banks or distance working at home are invading private life.

If these events are occurring, it is not by some insidious manipulation of the passive consumer but because the majority of citizens participate in the same "utopia of a better society through technology." This social imagination continues to mobilize the energy of some, as witnessed by the fact that hopes and fears are reborn as soon as an innovation is newly proposed for social use: In becoming an everyday technology, the

TABLE 3 Television Audience

Country	*Percent of Viewers Weekly*	*Minutes per Day*	*Percent of Housing with Cable/Shared Antenna*
U.S.A.	99	270	45/–
United Kingdom	79	228	1/5
Spain	87	207	–/–
Japan	88	190	13/–
France	86	178	2/–
Netherlands	80	140	51/23
Germany	71	137	6/40
Belgium*	72/84	136/126	83/–
Italy	80	129	–/–
Norway	80	124	12/19
Sweden	77	105	2/41
Ireland	80	102	35/–
Denmark	76	92	7/40

Includes all television stations (or only national TV for Denmark and Ireland).

*Belgium: French/Flemish

Source: Euro-fact book, basic hardware, and audience data, 1986–87 edition, O.E.R.F. Mediaforschung Wien.

innovation will participate in the construction of the utopian New World. This imagination gives meaning to the action of workers, citizens, consumers—whether they fight for or against the realization of this utopia.

Such an evolution signifies an indisputable and long-term reinforcement of the *techno-poliste* society. Within this familiar frame, a brief mark remains: Will people become "digital," eminently rational and quantitative in their representation of the physical world and others? There is no sure answer, due to a lack of perspective in time and in-depth research. The social imagination, however, guides our actions in this direction.

Notes

[1]Pierre-Alain Mercier, François Plassard, and Victor Scardigli, *La société digitale* (Paris: Editions du Seuil, 1984).

[2]Marc U. Porat, *The Information Economy* (Stanford University: Center for Interdisciplinary Research, 1976).

[3]Daniel Bell, "The Social Framework of the Information Society," in M. L. Dertouzos and J. Moses, eds., *The Computer Age* (Cambridge, Mass.: MIT Press, 1979).

[4]Yoneiji Masuda, *The Information Society as Post-Industrial Society* (Tokyo: Institute for the Information Society, 1980).

[5]Victor Scardigli, *La consommation culture du quotidien* (Paris: PUF, 1983).

[6]Victor Scardigli, "Vivre demain à technopolis," *Autrement*, 74 (Paris: November 1985), 247–51.

[7]Simon Nora and Alain Minc, *L'informatisation de la société* (Paris: La Documentation Française, 1978).

[8]Alvin Toffler, *La troisième vague* (Paris: Denoël, 1980).

[9]Radovan Richta, ed., *La civilisation au carréfour* (Paris: Editions du Seuil, 1974).

[10]Philippe Breton, *Histoire de l'informatique* (Paris: Editions de la Découverte, 1987).

[11]Mercier et al., *La société digitale.*

[12]Jean Baudrillard, *Simulacres et simulations* (Paris: Galilée, 1982).

[13]Marc Guillaume, "Télé-spectres," *Traverses*, 26 (1982), 18–28.

[14]Lyotard, Jean-François, trans. Geoff Bennington and Brian Massumi, *The Post-modern Condition: A Report on Knowledge*. Minneapolis: University of Minnesota Press, 1984 (Paris: Editions de minuit, 1979).

[15]Alain Gras, "Le pilote, l'écran et la secrétaire," *futuribles*, 104 (Paris: November 1986), 37–48.

[16]André Leroi-Gourhan's oral communication, bringing up to date his book *Le geste et la parole* (Paris: Albin Michel, 1965).

[17]Seymour Papert, *Mindstorms: Children, Computers and Powerful Ideas*, New York, Basic Books, 1980.

[18]Pierre-Alain Mercier, Victor Scardigli, and Roland Tourreau, *Télématique et vie locale* (Paris: CNRS-MCD, Report, 1986), p. 382.

[19]Claire Ancelin and Marie Marchand, *Promenade dans les usages* (Paris: La Documentation Française, 1984).

Bibliographical Essay

One of the main, explicit intentions of the book is to contribute, among other things, to the reorganization of existing theories and knowledge in a turbulent period of paradigmatic shifts. Given the varied horizons and disciplines represented by the authors, not to mention the diverse geographical origins, to give appropriate bibliographical information would have meant filling a whole book in fine print with a very limited utility for the already puzzled reader: Completedness would have killed usefulness. Instead, this bibliographical essay is intended as a personal, and probably biased, selection of "conceptual tools" that have never been very far away while preparing the book. Current technological, social, and intellectual transformations are represented, some might even think overrepresented, because in my opinion they must occupy an important space in design-related topics.

Agassi, Joseph, *Technology: Philosophical and Social Aspects*. Dordrecht: D. Reidel, 1985.

Aron, Jean-Paul, *Les modernes*. Paris: Gallimard, 1984.

Banham, Reyner, *Theory and Design in the First Machine Age*. Cambridge, Mass: MIT Press, 1980.

Barthes, Roland, *Mythologies*. Paris: Editions du Seuil, 1957. (trans. Annette Lavers, New York: Hill & Wang, 1972)

Bateson, Gregory, *Steps to an Ecology of Mind*. Novato, Calif.: Chandler and Sharps, 1972.

Baudrillard, Jean, *Pour une critique de l'économie politique du signe*. Paris: Gallimard, 1968. (trans. Charles Levin, New York: Telos 1978)

Baudrillard, Jean, *Le système des objets*. Paris: Denoël/Gonthier, 1968.

Baudrillard, Jean, *Le miroir de la production*. Paris: Castermann, 1973 (trans. Mark Poster, New York: Telos Press, 1975).

Baudrillard, Jean, *A l'ombre des majorités silencieuses, ou la fin du social*. Paris: Cahiers d'Utopie, 1978.

Beaune, Jean-Claude, *La technologie introuvable. Recherche sur la définition et l'unité de la technologie à partir de quelques modèles du XVIII et XIX Siècles*. Paris: Librairie philosophique J. Vrin, 1980.

Bell, Daniel, *The End of Ideology. On the Exhaustion of Political Ideas in the Fifties*. New York: Free Press, 1960.

Bell, Daniel, "Notes on the Post-Industrial Society," in *The Public Interest*, 6 (Winter 1967).

Bell, Daniel, *The Coming of Post-Industrial Society. A Venture in Social Forecasting*. New York: Basic Books, 1973.

Bell, Daniel, *The Cultural Contradictions of Capitalism*. New York: Basic Books, 1976.

Benjamin, Walter, *Einbahnstrasse* (1928). Frankfurt am Main: Suhrkamp, 1955.

Benjamin, Walter, *Das Passagen-Werk*, 2 volumes. Frankfurt am Main: Suhrkamp, 1982.

Berger, Peter L., *Facing Up to Modernity. Excursions in Society, Politics, and Religion*. New York: Penguin Books, 1977.

Berman, Marshall, *All That Is Solid Melts into Air. The Experience of Modernity*. New York: Simon & Schuster, 1982.

Bocchi, Gianluca, and **Mauro Ceruti**, eds., *La sfida della complessità*. Milan: Feltrinelli, 1985.

Boden, Margaret A., *Artificial Intelligence and Natural Man*. New York: Basic Books, 1981.

Bolter, David J., *Turing's Man. Western Culture in the Computer Age*. Chapel Hill: University of North Carolina Press, 1984.

Branzi, Andrea, *The Hot House: Italian New Wave Design*. Cambridge, Mass. MIT Press, 1984 (preface by Arata Isozaki).

Burnham, David, *The Rise of the Computer State*. New York: Random House, 1983.

Capra, Fritjof, *The Turning Point. Science, Society and the Rising Culture*. New York: Simon & Schuster, 1982.

Cervellati, Pier Luigi, *La città post-industriale*. Bologna: Il Mulino, 1984.

Ciribini, Giuseppe, *Introduzione alla tecnologia del design*. Milan: Franco Angeli, 1979.

Ciribini, Giuseppe, *Tecnologia e progetto*. Turin: Celid., 1984.

Dagognet, François, *Rematérialiser*. Paris: Librairie philosophique J. Vrin, 1985.

De Fusco, Renato, *Storia del design*. Bari: Laterza, 1985.

Derrida, Jacques, The Post Card: From Socrates to Freud and Beyond. *La carte postale*. Paris: Flammarion, 1979. (trans. Alan Bass, Chicago, University of Chicago Press, 1987).

Dertouzos, Michael L., and **Joel Moses**, eds., *The Computer Age: A Twenty-Year View*. Cambridge, Mass.: MIT Press, 1979.

De Sola, Pool, *Technologies of Freedom: on Free Speech in an Electronic Age*. Cambridge, Mass.: Harvard University Press, 1983.

Diebold, John, *Automation: The Advent of the Automatic Factory*. New York: Van Nostrand, 1952.

Dreyfus, Hubert, and **Stuart Dreyfus**, *Mind over Machine*. New York: Macmillan/Free Press, 1986.

Ellul, Jacques, *Le Système technicien*. Paris: Calmann/Lévy, 1977.

Elster, Jon, *Ulysses and the Sirens*. New York: Cambridge University Press, 1977.

Feigenbaum, Edward A., and **Pamela McCorduck**, *The Fifth Generation: Artificial Intelligence and Japan's Computer Challenge to the World*. Reading, Mass.: Addison-Wesley, 1983 and London: Michael Joseph, 1984.

Forester, Tom, ed., *The Microelectronics Revolution*. Cambridge, Mass.: MIT Press, 1981.

Forester, Tom, ed., *The Information Technology Revolution*. Cambridge, Mass.: MIT Press, 1985.

Forester, Tom, *High-Tech Society*. Cambridge, Mass.: MIT Press, 1987.

Forty, Adrian, *Objects of Desire: Design and Society 1750–1980*. New York: Pantheon, 1986.

Foster, Hal, ed., *The Anti-Aesthetic. Essays on Postmodern Culture*. Port Townsend, Wa.: Bay Press, 1983.

Frateili, Enzo, *Il disegno industriale italiano 1928–1981*. Turin: CELID, 1983.

Frude, Neil, *The Intimate Machine: Close Encounters with the New Computers*. London: Century, 1983 and New York: New American Library, 1983.

Gablik, Suzi, *Has Modernism Failed*? New York: Thames and Hudson, 1984.

Gaudin, Thierry, *A l'écoute des silences*. Paris: Gallimard, 1979.

Gershuny, Jonathan I., "Post-Industrial Society. The Myth of the Service Economy," in *Futures*, IX, no. 2 (April 1977).

Gershuny, Jonathan I., *After Industrial Society*? London: Macmillan, 1978.

Giedion, Siegfried, *Mechanization Takes Command*. New York: Oxford University Press, 1948.

Gille, Bertrand, "Prolégomènes à une histoire des techniques," in *Histoire des techniques*. Paris: Gallimard, 1978.

Gottmann, Jean, *Megalopolis. The Urbanized Northeastern Seaboard of the United States*. New York: Twentieth Century Fund, 1961.

Habermas, Jürgen, ed., *Stichworte zur "Geistigen Situation der Zeit," II, Politik und Kultur*. Frankfurt am Main: Suhrkamp, 1979.

Habermas, Jürgen, The Philosophical Discourse of Modernity (*Der philosophische Diskurs der Moderne*). Frankfurt am Main: Suhrkamp, 1985. (trans. Frederick G Lawrence, Cambridge, Mass., MIT Press)

Habermas, Jürgen, and **Niklas Luhmann**, *Theorie der Gesellschaft oder Sozialtechnologie*. Frankfurt am Main: Suhrkamp, 1971.

Handy, Charles, *The Future of Work*. Oxford and New York: Basil Blackwell, 1984.

Hassan, Ihab, and **Hassan, Sally**, eds., *Innovation/Renovation. New Perspectives on the Humanities*. Madison: University of Wisconsin Press, 1983.

Hirschheim, R. A., *Office Automation: A Social and Organizational Perspective*. Chichester, England and New York: Wiley, 1985.

Hofstadter, Douglas R., *Gödel, Escher, Bach. An Eternal Golden Braid*. New York: Basic Books, 1979.

"Informatique matin, midi . . . et soir!" *Autrement*, 37 (February 1982).

Informatique, télématique et vie quotidienne. Proceedings of the international symposium Informatique et société, 3 volumes. Paris: La Documentation Française, 1980.

Karl, Frederick R., *Modern and Modernism*. New York: Atheneum, 1985.

Kidder, Tracy, *The Soul of a New Machine*. Boston: Little, Brown, 1981 and Hardmondsworth, England: Penguin Books, 1982.

Klotz, Heinrich, ed., *Die Revision der Moderne. Postmoderne Architektur 1960–1980*. Munich: Prestel, 1986. (trans.

Kolb, David, *The Critique of Pure Modernity. Hegel, Heidegger and After*. Chicago: University of Chicago Press, 1986.

Krauss, Rosalind E., *The Originality of the Avant-Garde and other Modernist Myths*. Cambridge, Mass.: MIT Press, 1986.

Larroche, Hélène, and **Yan Tucny**, *L'objet industriel en question*. Paris: Editions du Regard, 1985.

Lyotard, Jean-François, The Post-Modern Condition: A Report on Knowledge. *La Condition post-moderne*. Paris: Editions de Minuit, 1979 (trans. Geoff Bennington and Brian Massumi, University of Minnesota Press, 1984).

Lyotard, Jean-François, *Tombeau de l'intellectuel et autres papiers*. Paris: Galilée, 1984.

Lyotard, Jean-François, *Les immatériaux*. Paris: Editions du Centre Pompidou, 1985.

Lyotard, Jean-François, *Le post-moderne expliqué aux enfants*. Paris: Galilée, 1986.

McLean, Mick, ed., *The Information Explosion. The New Electronic Media in Japan and Europe*. London: Frances Finter, 1985.

Maldonado, Tomás, *Beitrag zur Terminologie der Semiotik*. Ulm: Korrelat, 1961.

Maldonado, Tomás, *La speranza progettuale*. Turin: Einaudi, 1970.

Maldonado, Tomás, *Avanguardia e razionalità*. Turin: Einaudi, 1974.

Maldonado, Tomás, ed., *Tecnica e cultura. Il dibattito tedesco fra Bismark e Weimar*. Milan: Feltrinelli, 1979.

Maldonado, Tomás, *Disegno industriale: un riesame*. Milan: Feltrinelli, 1986.

Mandelbrot, Benoît, *Les objets fractales. Formes, hasard et dimension*. Paris: Flammarion, 1984.

Manzini, Ezio, *The Material of Invention. Materials and Design*. Cambridge, Mass.: MIT Press, 1988.

Margolin, Victor, ed., *Design Discourse. History Theory Criticism*, Chicago, University of Chicago Press, 1989

Marstrand, Pauline, ed., *New Technology and the Future of Work and Skills*. London and Dover, N.H.: Frances Pinter, 1984.

Martin, James, and **Adrian R. D. Norman**, *The Computerized Society*. Englewood Cliffs, N.J.: Prentice-Hall, 1970.

Martin, James, *Telematic Society. A Challenge for Tomorrow*. Englewood Cliffs, N.J.: Prentice-Hall, 1981.

Mitcham, Carl, and **Alois Huning**, eds., *Philosophy and Technology II. Information Technology and Computers in Theory and Practice*. Dordrecht: D. Reidel, 1986.

Moles, Abraham, *Théorie de l'information et perception esthétique*. Paris: Flammarion, 1958.

Moles, Araham, *Sociodynamique de la culture*. Paris: Mouton, 1967.

Moles, Abraham, "Objet et communication," in *Communications*, 13 (1969).

Moles, Abraham, and **Elisabeth Rohmer**, *Psychologie de l'espace*. Paris: Casterman, 1972.

Moore, Gary T., ed., *Emerging Methods in Environmental Design and Planning*. Cambridge, Mass.: MIT Press, 1970.

Moos, Stanislaus von, and **Chris Smeenk**, eds., *Advantgarde und Industrie*. Delft: Delft University Press, 1983.

Morello, Augusto, and **Anna Castelli Ferrieri**, *Plastiche e design*. Milan: Arcadia, 1984.

Morin, Edgar, *Pour sortir du vingtième siècle*. Paris: Nathan, 1981.

Moulakis, Athanasios, ed., *The Promise of History*. Berlin: W. De Gruyter, 1985.

Naisbitt, John, *Megatrends*. New York: Warner Books, 1982.

Newman, Charles, *The Post-Modern Aura: the Act of Fiction in an Age of Inflation*. Evanston, Ill.: Northwestern University Press, 1984.

Nikerson, Raymond S., *Using Computers: The Human Factors of Information Systems*. Cambridge, Mass.: Bradford Books/MIT Press, 1986.

"1984" et les présents de l'univers informationnel. Paris: Centre Georges Pompidou/Centre de Création Industrielle, 1985.

Nora, Simon, and **Alain Minc**, *L'informatisation de la société*. Paris: La Documentation Française, 1978.

Papert, Seymour, *Mindstorms: Children, Computers and Powerful Ideas*. New York: Basic Books, 1980.

Pardi, Francesco, and **Giovan Francesco Lanzara**, *L'interpretazione della complessità*. Naples: Guida, 1980.

Polanyi, Karl, *The Great Transformation*. New York: Holt, Rinehart & Winston, 1944.

Prigogine, Ilya, and **Isabelle Stengers**, *La nouvelle alliance, metamorphose de la science*. Paris: Gallimard, 1979.

Quiniou, Jean-Claude, *Télématique, mythes et réalités*. Paris: Gallimard, 1980.

Rajchman, John, and **Cornel West**, eds., *Post-Analytical Philosophy*. New York: Columbia University Press, 1985.

Robertson, James, *Future Work: Jobs, Self-Employment and Leisure after the Industrial Age*. Aldershot, England: Gower, 1985.

Roqueplo, Philippe, *Penser la technique*. Paris: Editions du Seuil, 1983.

Roqueplo, Philippe, and **Jacques Zeitoun**, *Cultiver la technologie. Création et technologies*. Paris: Dalloz, 1983.

Rose, Frank, *Into the Heart of the Mind: A Quest for Artificial Intelligence*. Boston: Little, Brown, 1981 and Hardmondsworth, England: Penguin Books, 1982.

Rosenberg, Bernard, and **David Manning White**, eds., *Mass Culture*. Glencoe, Ill.: Free Press, 1957.

Roszak, Theodore, *The Cult of Information. The Folklore of Computers and the True Art of Thinking*. New York: Pantheon Books, 1986.

Ruberti, Antonio, ed., *Technologia domani. Utopie differite e transizioni in atto*. Bari: Laterza, 1985.

Salomon, Jean-Jacques, *Prométhée enchaîné. La résistance au changement technique*. Paris: Pergamon Press, 1981.

Santachiara, Denis, ed., *Le néomerce*. Milan: Electa/Triennale di Milano, 1985.

Schoeler, Andreas von, ed., *Informationsgessellschaft oder überwachungsstaat?* Opladen: Westdeutscher, 1986.

Searle, John R., *Speech Acts*. Cambridge: Cambridge University Press, 1969.

Searle, John R., *Expression and Meaning. Studies in the Theory of Speech Acts*. Cambridge: Cambridge University Press, 1979.

Shaiken, Harley, *Work Transformed: Automation and Labor in the Computer Age*. New York: Holt, Rinehart & Winston, 1985.

Simondon, Gilbert, *Du mode d'existence des objets techniques*. Paris: Aubier Montaigne, 1958.

Simon, Herbert A., *Models of Man*. New York: Wiley, 1957.

Simon, Herbert A., *The Science of the Artificial*. Cambridge, Mass.: MIT Press, 1969.

Sparke, Penny, Felice Hodges, Emma Dent Coard, and Anne Stone, *Design Source Book: A Visual Reference to Design from 1850 to the Present Day*. London: MacDonald, 1986.

Strassman, Paul A., *Information Payoff: The Transformation of Work in the Electronic Age*. New York: Free Press, 1985.

Tarde, Gabriel, *Les lois de l'imitation*. Paris: Alcan, 1890.

Thackara, John, ed., *Design After Modernism*, New York, Thames and Huolson, 1989.

Théofilakis, Elie, ed., *Modernes, et après? Les immatériaux*. Paris: Autrement, 1985.

Toffler, Alvin, *Future Shock*. New York: Random House, 1970.

Toffler, Alvin, *The Third Wave*. New York: Morrow, 1980.

Toulmin, Stephen, and **June Goodfield**, *The Discovery of Time*. Chicago: University of Chicago Press, 1982.

Touraine, Alain, *La société postindustrielle*. Paris: Editions Denoël, 1969.

Trini Castelli, Clino, *Il lingotto primario*. Milan: Arcadia, 1985.

Vattimo, Gianni, *La fine della modernità*. Milan: Garzanti, 1985.

Vattimo, Gianni, and **Pier Aldo Rovatti**, eds., *Il pensiero debole*. Milan: Feltrinelli, 1983.

Venturi, Robert, *Complexity and Contradiction in Architecture*. New York: Museum of Modern Art, 1977.

Virilio, Paul, *Vitesse et politique*. Paris: Galilée, 1977.

Virilio, Paul, *L'espace critique*. Paris: Christian Bourgois, 1984.

Warrick, Patricia S., *The Cybernetic Imagination in Science Fiction*. Cambridge, Mass.: MIT Press, 1980.

Weizenbaum, Joseph, *Computer Power and Human Reason: From Judgment to Calculation*. San Francisco: Freeman, 1976 and London and New York: Penguin Books, 1984.

Wiener, Norbert, *The Human Use of Human Beings*. New York: Houghton Mifflin, 1950.

Winograd, Terry, and **Fernando Flores**, *Understanding Computers and Cognition*. Norwood, N.J.: Ablex, 1986.

Wolfe, Tom, *From Bauhaus to Our House*. New York: Farrar Straus Giroux, 1981.

Woodward, K., ed., *The Myths of Information: Technology and Post-Industrial Culture*. Madison, Wis.: Coda Press, 1980.

Index